W9-DDS-587

NASA & THE EXPLORATION OF SPACE

BY ROGER D. LAUNIUS
NASA CHIEF HISTORIAN

AND BERTRAM ULRICH
CURATOR, NASA ART PROGRAM

FOREWORD BY
SENATOR JOHN GLENN

THE

WITH WORKS FROM

NASA & EXPLORATION OF SPACE

THE NASA ART COLLECTION

STEWART, TABORI & CHANG
NEW YORK

DEDICATION

This book is dedicated to Susan Lawson-Bell, former Art Curator of the National Air and Space Museum—who conceived the idea of this project—as well as the original curators of the NASA Art Program, James Dean and Robert Schulman. All three fostered art to retell NASA's story.

PAGES 2–3:
Go for the Stars
P. A. NISBET
A NASA spacecraft blasts off into space.

PAGES 4–5:
Traveler on the Voyage
ROBERT A.M. STEPHENS
An imaginery artist-as-astronaut renders the space shuttle in orbit.

PAGES 6–7:
Saturn Blockhouse
FRED FREEMAN
The Saturn Blockhouse during launch (with artist).

PAGES 8–9:
Cape Winds
ATTILA HEJJA
The space shuttle is rolled out to the launch pad on a giant crawler.

PAGE 12:
Splashdown
ROBERT MCCALL
The Apollo 11 crew successfully splashed down into the Pacific Ocean on July 24, 1969.

Text copyright © 1998 Roger D. Launius
All works copyright © 1998 NASA Art Program or the National Air and Space Museum, except the following:

Sky Garden, by Robert Rauschenberg © 1969, Gemini G.E.L., Los Angeles.

Suit Up, Astronaut on the Moon, and *Behind the Apollo 11* reproduced by permission of the Norman Rockwell Family Trust.

Peter Hurd quote originally published in *Art in America,* Brant Publications, Inc., 1963.

Publisher: Lena Tabori
Project Director: Charles O. Hyman
Project Editor: Mary Kalamaras
Designer: Nai Y. Chang
Art Curator: Bertram Ulrich
Production: Deirdre Duggan

All Rights Reserved. No part of this publication may be reproduced, stored in a retrieval system, or transmitted in any form or by any means, electronic, mechanical, photocopying, recording or otherwise without written permission from the publisher.

Published in 1998 and distributed in the U.S. by
Stewart, Tabori & Chang,
a division of U.S. Media Holdings, Inc.
115 West 18th Street, New York, NY 10011

Distributed in Canada by
General Publishing Company Ltd.
30 Lesmill Road
Don Mills, Ontario, Canada M3B 2T6

Sold in Australia by
Peribo Pty Ltd.
58 Beaumont Road
Mount Kuring-gai, NSW 2080, Australia

Distributed in all other territories by
Grantham Book Services Ltd.
Isaac Newton Way, Alma Park Industrial Estate
Grantham, Lincolnshire, NG31 9SD, England

Library of Congress Cataloging-in-Publication Data

Launius, Roger D.
NASA & the exploration of space : with works from the NASA art collection / text by Roger D. Launius ; foreword by John Glenn.
p. cm.
Includes bibliographical references and index.
ISBN: 1-55670-696-0
1. Astronautics—United States—History. 2. United States. National Aeronautics and Space Administration—History. 3. Outer Space—Exploration—United States—History. 4. Outer space—In art. 5. Space vehicles in art. 6. Space ships in art. I. Title. II. Title: NASA and the exploration of space.
TL789.8.U5L38 1998
629.4'0973—DC21 98-17742
CIP

Printed in Japan by Toppan Printing Co.

10 9 8 7 6 5 4 3 2 1

The Artist and the Space Shuttle
Henry Casselli
NASA Artist Wilson Hurley is seen painting his impression of a shuttle launch.

"Space is man's great frontier to explore, and if history shows us anything, the artist should be a witness to our triumphs and tragedies along the way. . . . Aerospace exploration, although fraught with specialized machinery and awesome natural dangers, is no less a dream shared by all Americans than the road to cross the Great Divide and settle the Badlands. Artists have made those journeys and should make this one as well. It is the artist more than anyone else who can show us the 'color' of darkness, the 'weight' of weightlessness, the 'shape' of infinity."

Artist Chet Jezierski

CONTENTS

FOREWORD

Untitled
J. N. ROSENBERG
A festive parade in honor of John Glenn after his historic mission in 1962.

Americans are by their nature a questing people. We've been this way throughout our history. From the early settlers' struggle to survive on our rocky shores and Lewis and Clark's push into hostile lands west of the Mississippi to the Wright Brothers straining to break the bond of gravity and the past and present day pioneers of our space program, we have always sought to explore the unknown and to push the boundaries of what previously had been thought impossible. And in the effort to discover what's out there, we've learned more about ourselves than we ever could have imagined.

This questing nature is one of the best features of the American spirit. It spawns optimism and energy, curiosity and problem solving, and a willingness and determination to press on no matter how difficult the challenge or obstacle. We see it everyday—in the entrepreneur starting a new business, the student tackling a new subject, or the scientist discovering a technological breakthrough. No place is it more evident than in the National Aeronautics and Space Administration and the men and women who work there.

Through stunning visual imagery and informative text, *NASA & The Exploration of Space* captures the mood, feeling, and history of our space program as it reaches its 40th anniversary. It also highlights NASA's historical and art programs, and their important but sometimes overlooked contributors in our experiment in space.

The NASA Art Program provides small stipends to artists for pieces that emphasize different aspects of our space journey. It has funded works by the well-known—Robert Rauschenberg, Norman Rockwell, and James Wyeth among them—and the lesser known. Whether depicting a sense of the grand scale of technological brilliance, the beauty of the outer world, or the many emotions of man, the work within this book portrays our space program in its fullest scope, including its highs, lows, and many in-betweens.

Accompanying these images is a thorough narrative by Dr. Roger Launius, NASA's chief historian. Dr. Launius provides an excellent overall primer on the history of space exploration examining past accomplishments and current undertakings, as well as what can be expected in the future. Reading the first few sections followed by Chapter Three: From Competition to Cooperation, Launius's description of the evolution of the space relationship between the U.S. and the Soviet Union leaves me filled with a sense of irony. For those of us who entered the program with the cry of "the Russians are coming" and the beep of Sputnik ringing in our ears, the fact that we are now building together with our former adversaries a permanent facility for people to live and work in space was something beyond comprehension.

We've had fabulous accomplishments in forty years that are too many for me to list but are fully portrayed in the pages that follow. And I believe that our quest is still in its infancy.

Most importantly, it's all for the benefit of people on Earth. The struggle to explore beyond our planet and push human potential to its fullest will yield benefits that we cannot yet grasp, but our children and their children after them will one day realize them. Our legacy will be marked by not only the many tangible benefits of the space program, but by our ability to instill that questing nature of the American spirit in the generations to come.

SENATOR JOHN GLENN
MARCH 31, 1998

INTRODUCTION: A PREHISTORY OF THE SPACE AGE

STS-4 at 0052 Seconds.
Ren Wicks
Photographers and press watch the June 27, 1982, launch of the Space Shuttle Columbia.

There have been, according to historian Stephen Pyne, three great ages of exploration in the history of Western civilization, each of which has profoundly shaped the worldview of its respective population. The first, brought on by the Renaissance, was characterized by the shipboard explorations by such individuals as Columbus and Magellan, which resulted in a general mapping of the size and extent of the continents of Earth, as well as a general European expansion of influence over other territories. The second age, which overlapped the first in places and circumstances between the sixteenth and nineteenth centuries, was dominated by land expeditions by European adventurers such as Coronado, La Salle, Lewis and Clark, Stanley and Livingston, and others, who filled in many of the details of the continental interiors.

These eras of discovery could not have come about without a corresponding scientific basis of knowledge and technology that could allow explorers to succeed. The building of such a storehouse of knowledge required the support of several factors, among them a strong economic base and stable internal political environment and, above all, the political will and the public perception that explorations were both achievable and desirable. This is no less true with regard to the third and present age of exploration defined by the quest into regions inhospitable to humankind, most notably the North and South poles, the great ocean depths, and the frontiers of space. Of these three areas, however, it is the exploration of space that has had perhaps the greatest impact on twentieth-century Western civilization.

On September 12, 1963, President John F. Kennedy recognized this great third age when he spoke of the prospect of exploring space to a packed audience at the United Nations. Kennedy told of the great British explorer George Mallory who, when asked why he wanted to climb Mount Everest, simply stated, "Because it was there." The president then remarked, "Well, space is there, and we're going to climb it. And the Moon and the planets are there and new hopes for knowledge and peace are there. And therefore, as we set sail, we ask God's blessing on the most hazardous and dangerous and greatest adventure on which man has ever embarked."

Kennedy's words captured the excitement of an entire people and are no less poignant more than a generation later. Space exploration, although still in its infancy, has already changed our lives in fundamental ways. The modern world and the manner in which people live their lives today would be vastly different without the satellite telecommunications and meteorology and other advances in technology such exploration has fostered, including the present knowledge of the Moon and the planets, greater comprehension of the cosmos and the origins of the Universe, the Earth resource satellites that have enhanced our understanding of climate changes such as El Niño and ozone depletion, and the perspective of Earth from space and its impact on ecological conservation.

Like the eras that have preceded it, the age of space exploration has also managed to capture the momentum and excitement for geographical expansion, impacting profoundly upon many in Western civilization. From the earliest experimental rockets to the most sophisticated shuttle missions, awe, wonder, and hope have always ridden along. Writer Ray Bradbury once said that experiencing the launch of the Saturn V rocket for a flight to the Moon during Project Apollo was one of the most moving experiences of his life: "It shakes the rust from your body and connects you at once with the glory of the Universe." Rocket launches, no less dramatic now with the space shuttle than they were with the Saturns, are only the first step off this planet as humanity explores the space beyond. To understand how we have arrived at this third great era of exploration, we need to examine the events, both successful and challenging, that took place, as well as the visionaries who contributed to what has become the most daring and promising ages of exploration.

THE BIRTH OF MODERN ROCKETRY

The possibility of actually going into space and exploring it firsthand using individuals and machines began near the turn of the century with the advent of rocket technology. The origin of this scientific field resulted largely from the efforts of a small group of pioneers who developed the theoretical underpinnings and the technical capabilities of rocketry in the first half of this century: Konstantin E. Tsiolkovskiy, Hermann Oberth, Robert H. Goddard, and Robert Esnault-Pelterie. These men, considered the godfathers of modern space exploration, laid the fundamental groundwork for space flight with their experiments, and they inspired others to follow in their footsteps.

Perhaps the greatest of these pioneers, however, was Robert H. Goddard, considered the Father of Modern Rockery. He was born in Worcester, Massachussetts, on October 5, 1882, the son of a machine-shop owner. After completing his education, Goddard turned his attention to liquid rocket propulsion, publishing a classic study, *A Method of Reaching Extreme Altitudes,* under the auspices of the Smithsonian Institution in 1919. In it, Goddard argued from a firm theoretical base that rockets could be used to explore the upper atmosphere. Moreover, he suggested that with a velocity of 6.95 miles per second, without air resistance, an object could escape Earth's gravity and head into infinity, or to other celestial bodies. This became known as the Earth's "escape velocity" and Goddard believed that by using his theories and techniques, humans could reach the Moon.

Unfortunately, Goddard's ideas were considered a great joke by those who believed space flight was either impossible or impractical, much to the consternation of the already shy Goddard. The popular press was a source of ridicule; *The New York Times* was especially harsh in its criticisms, referring to him as a dreamer whose ideas had no scientific validity and questioning both Goddard's credentials as a scientist and the Smithsonian's rationale for funding his research and publishing his results.

This negative press essentially forced Goddard to go underground with his research. His quiet, isolated work on the design of liquid-fueled rockets resulted in the launching of history's first successful rocket on March 16, 1926, near Auburn, Massachusetts. Although it rose only 184 feet, the feat heralded the modern age of rocketry. Despite this success, Goddard's continued experiments drew increasing public notice and, along with it, additional media mockery. He decided to seek relief and solitude in the American Southwest.

Fortunately, Charles A. Lindbergh, fresh from his transatlantic solo flight, became interested in Goddard's work. He visited Goddard and was sufficiently impressed to persuade philanthropist Daniel Guggenheim to award Goddard a $50,000 grant. Goddard used the grant to set up an experiment station in a desolate area near Roswell, New Mexico, where between 1930 and 1941 he launched larger rockets of increasingly greater complexity and capability. In addition, he developed steering systems for in-flight rockets, which incorporated a rudderlike device to deflect the gaseous exhaust and gyroscopes to keep the rocket headed in the proper direction.

In late 1941, Goddard entered the naval service and spent World War II working on rocket engines, among them the throttleable Curtiss-Wright XLR25-CW-1 rocket engine that later powered the Bell X-1 and helped overcome the transonic barrier in 1947. Sadly, Goddard did not live to see this; he died in Baltimore, Maryland, on August 10, 1945.

CALTECH AND THE EARLY JPL

During the time of Goddard's research into liquid-fueled rockets, related scientific activities were also taking place at the Guggenheim Aeronautical Laboratory of the California Institute of Technology in Pasadena, California, the results of which became more widely disseminated than Goddard's work and, perhaps for this reason, were considered more significant in the field of rocketry.

Frank J. Malina, a young, enthusiastic Caltech Ph.D. student, had adopted a research agenda for the design and experimentation of a high-altitude sounding rocket. Beginning in late 1936, Malina and his colleagues—a brilliant and capable rocket team—began static testing rocket engines in the Arroyo Seco Canyon above the Rose Bowl in Pasadena with mixed results. Calling themselves the "suicide squad," they took chances with these combustibles that would never be tolerated today. (One of the more eccentric of the rocketeers, a member of a magic cult, blew himself up by accidentally dropping a vial of fulminate of mercury.) Finally, on November 28, 1936, their motor ran for fifteen seconds. A series of tests thereafter brought incremental improvements; a year later Malina and an associate had learned enough to distill the results into the first scholarly paper on rocketry to come out of Caltech. The test results showed that with proper fuels and motor efficiency, a rocket could be constructed with the capability to ascend as high as 1,000 miles. As a result of this research, Malina's rocketry team obtained funding from the Army Air Corps to build rockets to assist aircraft takeoff. They began working in 1939 on what became the jet-assisted takeoff (JATO) project at what later came to be known as the Jet Propulsion Laboratory (JPL), a contractor facility operated by Caltech.

Because of the potential implications and applications of his groundbreaking research, Malina had always expressed misgivings about working on weaponry. He even went so far as to accept post–World War II employment with the United Nations to assist in preventing further violence. However, the difficult political climate in 1939 prompted him ultimately to support the buildup of U.S. military capability as a deterrent to fascism.

THE ROCKET AND MODERN WARFARE

During World War II, virtually every belligerent was involved in developing some type of rocket technology—the entry-level technology for space flight. Malina and his team were busy developing the WAC Corporal sounding rocket, a launch vehicle capable of sending a 25-pound payload to an altitude of 100,000 feet. First flown on October 11, 1945, the WAC Corporal became a significant launch vehicle in postwar rocket research.

Of the nations that were experimenting, however, Nazi Germany expended the most resources and enjoyed the most success. This was perhaps largely due to the efforts of the charismatic and politically astute twenty-year-old Wernher von Braun, hired by the German army in 1932 to work in its military rocket program. In retrospect, questions have risen over Von Braun's motivations for this move in light of Hitler's rise to power and the devastation and terror of World War II. For some, von Braun was a visionary who foresaw the potential of human space flight, but for others he was little more than an arms merchant who developed brutal weapons of mass destruction. In reality, he seems to have been something of both, never evincing Malina's hesitancy about the morality of using scientific and technical knowledge to kill as many people and destroy as many resources as possible.

On June 13, 1942, at a secret base at Peenemünde on the Baltic Coast, von Braun conducted the first test of the five-and-a-half-ton German A-4 (V-2) rocket. Although unsuccessful, another launch effort four months later

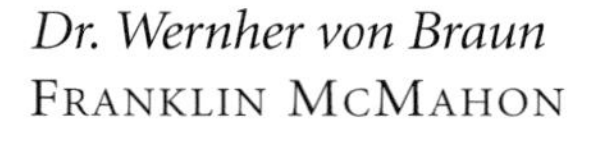

Dr. Wernher von Braun
FRANKLIN MCMAHON

resulted in the rocket traveling more than 120 miles. The first operational use of the rocket came in September 1944 when Germany employed V-2s against Allied targets in France, Belgium, and England. Traveling at speeds in excess of 3,500 miles per hour, the rockets delivered 2,200-pound warheads. The Germans also fired over one hundred V-2s from Blizna, Poland, launching ten on one day alone. At the time of German collapse, which formally took place on May 8, 1945, more than 20,000 V-weapons, V-1s and V-2s, had been fired—and with unreliable results; the majority of the rockets never reached their targets. Although figures vary, the best estimate is that of this total, 1,115 V-2 rockets were launched against England and 1,675 against continental targets such as Antwerp, Belgium.

POSTWAR ROCKET TECHNOLOGY AND SPACE SCIENCE

Although the V-2 was far from a decisive weapon for Germany—only the atom bomb deserves such a characterization—the United States wanted to learn as much about this technology as possible. Accordingly, in August 1945 the Army shipped recovered components for approximately one hundred V-2 ballistic missiles from Europe to the White Sands Proving Ground in New Mexico. Along with them—as part of a secret military operation called Project Paperclip—came many of the scientists and engineers who had developed these weapons, notably von Braun, who surrendered to the United States in hopes of continuing his rocketry experiments under U.S. sponsorship. Between 1946 and 1951, the United States spent significant sums to assemble and launch sixty-seven captured V-2s on nonorbital flights.

As one journalist commented in 1952, von Braun was "the major prophet and hero (or wild propagandist, some scientists suspect) of space travel" who lost World War II for Germany by diverting resources from useful activities into the V-2. In the case of rocket development in the United States in the latter 1940's, von Braun's presence took attention and resources away from the very real technical progress being made at JPL, as well as other places around the country that might have otherwise yielded earlier results.

It was not until the ballistic missile programs under the Eisenhower administration in the 1950's that rocketry made significant strides beyond World War II technology. The enormous advantages of the United States in technical capability that had been honed to a sharp edge in World War II were dismantled in the rush to demobilize in 1945. Most Americans, secure in their transitory nuclear dominance and thinking of national strategy only in terms of World War II models, did not appreciate the necessity of developing rockets for future national defense and space exploration.

With the rise of the Cold War during the 1950's, President Eisenhower set the armed services on a path toward fielding intercontinental ballistic missiles (ICBMs) that could deliver warheads to targets in the Soviet Union. The real ICBM break came in 1954 when Eisenhower accepted the recommendations by the advisory committee set up to assure the early success of the Atlas ICBM, a program that had languished throughout the Truman years. To accelerate this and other ICBM projects, the president established a Ballistic Missile Division under the leadership of Air Force Brigadier General Bernard A. Schriever. By the summer of 1955, the Air Force managed two accelerated ICBM projects (Atlas and Titan) and the Thor intermediate-range ballistic missile (IRBM) effort. The Army, not to be left out of the rush, approved the development of the Jupiter IRBM under the direction of Wernher von Braun. Although only a slight improvement over the V-2, it proved extremely reliable. The Navy also saw a role for ICBMs on its nuclear submarines and proceeded to build the solid-propellant Polaris missile. The Air Force later turned its attention to solid-fueled rockets and in 1957 began development of the Minuteman. By the fall of 1957 (the time of Sputnik), the United States military had under development six missiles on a "crash basis." All would provide important technologies for later space exploration, and the liquid-fueled Atlas, Titan, and Jupiter (especially its Redstone variant) would find direct application as launch vehicles for both satellites and astronauts.

The investment made by the U.S. Department of Defense (DOD) for this technology was tremendous. In the summer of 1957, the National Security Council (NSC) reported that the nation had spent $11.8 billion on military space activities. When it projected that ICBM capability would become a reality, the NSC determined that the cost of continuing the programs from 1957 through 1963 would be an additional $36 billion.

It should be noted that not all rocket developments were military in focus; many were conducted on a moderate schedule and without the funding available to DOD efforts. For example, the Aerobee, a scaled-up version of the WAC Corporal developed by JPL, launched a sizable payload to an altitude of 130 miles at a very economical cost. This reliable booster enjoyed a long career from its first instrumented firing on November 24, 1947, until the January 17, 1985, launch of the 1,037th and final Aerobee. Also, the Naval Research Laboratory, a scientific arm of the Navy, proposed to build a large sounding rocket, the Viking launch vehicle manufactured by the Glenn L. Martin Company. It first took off from White Sands on May 3, 1949; the twelfth and last one flew on February 4, 1955.

SPUTNIK NIGHT

The combination of technological and scientific advance, the thrill of exploration, political competition with the Soviet Union, and changes in popular opinion about space flight came together in a very specific way in the 1950's to affect public policy in favor of an aggressive space program. This found tangible expression in 1952 when the international scientific organization known as CSAGI (Comité Speciale de l'Année Geophysique Internationale) began planning for the International Geophysical Year (IGY), an international scientific research effort to study geophysical phenomena. It was decided that July 1, 1957, through December 31, 1958, would be the period of emphasis, in part because of a predicted expansion of solar activity.

In October 1954, the council met in Rome, Italy, and adopted a resolution calling for the launch during the IGY of artificial satellites that would help map the Earth's surface. Immediately, the Soviet Union announced its plan to orbit an IGY satellite, a move that would virtually assure a response from the United States. In 1955, the United States did answer with its own IGY scientific satellite program, Project Vanguard, to be headed by Dr. John P. Hagen, a senior scientist with the Naval Research Laboratory.

On September 30, 1957, the CSAGI opened a six-day conference at the

National Academy of Sciences in Washington on rocket and satellite research for the IGY. Scientists from the United States, the Soviet Union, and five other nations met to discuss their individual plans and to develop protocols for sharing scientific data and findings. During the conference several Soviet officials had intimated that they could probably launch their scientific satellite within weeks, instead of months as the public schedule had indicated. Hagen worried that scientist Sergei M. Poloskov's offhand remark made on the conference's first day—that the Soviet Union was "on the eve of the first artificial earth satellite"—was more than mere boastful rhetoric. Hagen was concerned over what a surprise Soviet launch would mean for his Vanguard program—already behind schedule and over budget—as well as for the United States itself.

Nevertheless, few Americans expected that the events during the evening reception held at the Soviet Embassy in Washington, D.C., the following Friday, October 4, would change the course of history and the path of the Cold War. Just before 6:00 P.M., Walter Sullivan, a *New York Times* reporter in attendance, received a frantic telephone call from his Washington bureau chief: *Tass* had just announced the launch of Sputnik 1, the world's first Earth-orbiting artificial satellite. Returning to the party, Sullivan sought out Richard Porter, a member of the American IGY committee, and whispered, "It's up." Porter then went in search of Lloyd Berkner, the official American delegate to CSAGI.

When told the news, Berkner, retaining his polished southern-gentleman demeanor, clapped his hands for attention and asked for silence. "I wish to make an announcement," he declared. "I've just been informed by the *New York Times* that a Russian satellite is in orbit at an elevation of nine hundred kilometers. I wish to congratulate our Soviet colleagues on their achievement." Hagen was shocked. The Soviets had beaten the Vanguard satellite effort into space. Scientists immediately adjourned to the embassy's rooftop to view the heavens. Although they were unable to see the satellite with the naked eye, Sputnik I had indeed twice passed within easy detection range of the United States before anyone knew of its existence.

The inner turmoil that Hagen felt on "Sputnik Night," as that evening came to be called, reflected the sentiment of the American public in the days that followed. Word today still cannot easily convey the American reaction to the Soviet satellite. The only characterization that comes close to capturing the mood of October 5, 1957, is hysteria. A collective mental turmoil and soul-searching followed, as American society thrashed around for the answers to Hagen's question: Were the Soviets really the greatest nation on Earth, as their leaders claimed? Almost immediately, the phrases pre-Sputnik and post-Sputnik entered the American lexicon to redefine time. However, with the launch of Sputnik I, a more profound term emerged to change the world thereafter: space age.

Launched from the Soviet Union's rocket-testing facility in the desert near Tyuratam in the Kazakh Republic, Sputnik proved a decidedly unspectacular satellite that probably should not have elicited the horrific reaction it wrought. An 22-inch aluminum sphere with four spring-loaded whip antennae trailing, it weighed only 183 pounds and traveled an elliptical orbit that took it around the Earth every 96 minutes. It carried a small radio beacon that beeped at regular intervals and could by means of telemetry verify exact

Baikonur: A Kazakh Sunrise
Alan Chinchar
Dawn splashes on the rails of the Biakonur Cosmodrome, the Russian launch complex.

locations on the Earth's surface. The satellite itself fell from orbit three months after launch on January 4, 1958.

While President Eisenhower and other leaders of his administration congratulated the Soviets and tried to downplay the importance of the accomplishment, they misjudged the public reaction to the event. The launch of Sputnik 1 had a "Pearl Harbor" effect on American public opinion. It was a shock, introducing the average citizen to the space age in a crisis setting. The event created the illusion of a technological gap and provided the impetus for increased spending for aerospace endeavors, technical and scientific educational programs, and the chartering of new federal agencies to manage air and space research and development. Not only had the Soviets been first in orbit, but Sputnik 1 was much larger than the intended three-and-a-half pound satellite to be launched in Project Vanguard. In the Cold War environment of the late 1950's, this disparity of capability held menacing implications.

Concerns were raised to an even higher level on November 3, 1957, less than a month later, when the Soviet Union struck again by launching Sputnik 2 carrying a dog, Laika. This spacecraft weighed 1,120 pounds and stayed in orbit for almost 200 days. The dog, although lost during the mission, verified the possibility of life surviving space flight.

THE AMERICAN RESPONSE

In the wake of the Sputnik crisis, Senate Majority Leader Lyndon B. Johnson, a Democrat from Texas, realized something had to be done. He opened hearings by a subcommittee of the Senate Armed Services Committee on November 25 to review the entire spectrum of American defense and space programs. This group found serious underfunding and incomprehensible organization for the conduct of space activities, for which it blamed the president and the Republican party.

Although Johnson's concern was in all probability legitimate, he had also recognized and exploited the political opportunity brought about by the Sputnik crisis and the resulting political debate. In the early years of the Cold War, the Republicans were the opposition party and in 1949 criticized President Harry S. Truman for the ouster of Chiang Kai-Shek in Nationalist China by the replacement communist government under Mao Zedong; Truman was again criticized in 1950 for the invasion and near capitulation to communist forces of South Korea. Wisconsin Republican Senator Joseph McCarthy had used these events to label the Democrats as soft on communism and of essentially allowing the "Red Menace" to conquer the world. The Republicans had turned these issues into political capital that swept Dwight D. Eisenhower into the presidency in 1952 along with a host of Republican members of Congress. Now the shoe was on the other foot, and the possibility existed of defeating the Republicans on the very same issue: Cold War rivalry.

Seeing this, the Eisenhower administration had to move quickly to restore confidence at home and prestige aboard. As the first tangible effort to counter the apparent Soviet leadership in space technology, the White House announced that the United States would test-launch a Project Vanguard booster on December 6, 1957. The media was invited to witness the launch in the hope that they could help restore public confidence, but the event was a disaster of the first order. During the ignition sequence the rocket rose about three feet above the platform, shook briefly, and disintegrated in flames. On February 5, 1958, the Vanguard launch vehicle reached an altitude of four miles and then exploded. John Hagen, who had been working feverishly to ready the rocket for flight, was demoralized, and some of his associates even felt that his career ended then and there, for he never again held an important post.

In this crisis, the Army's Wernher von Braun and his rocket team dusted off an unapproved plan for the IGY satellite effort, Project Explorer, and flew it within an amazingly short period of time. After two launch aborts that made observers nervous that the United States might never duplicate the Soviet successes in space flight, the Juno I booster carrying Explorer 1 lifted off from its Cape Canaveral, Florida, launch site at 10:55 P.M. on January 31, 1958. The tracking sites marked the course of the rocket to the upper reaches of the atmosphere and at 12:49 A.M., the JPL tracking station confirmed the passing of Explorer 1 overhead.

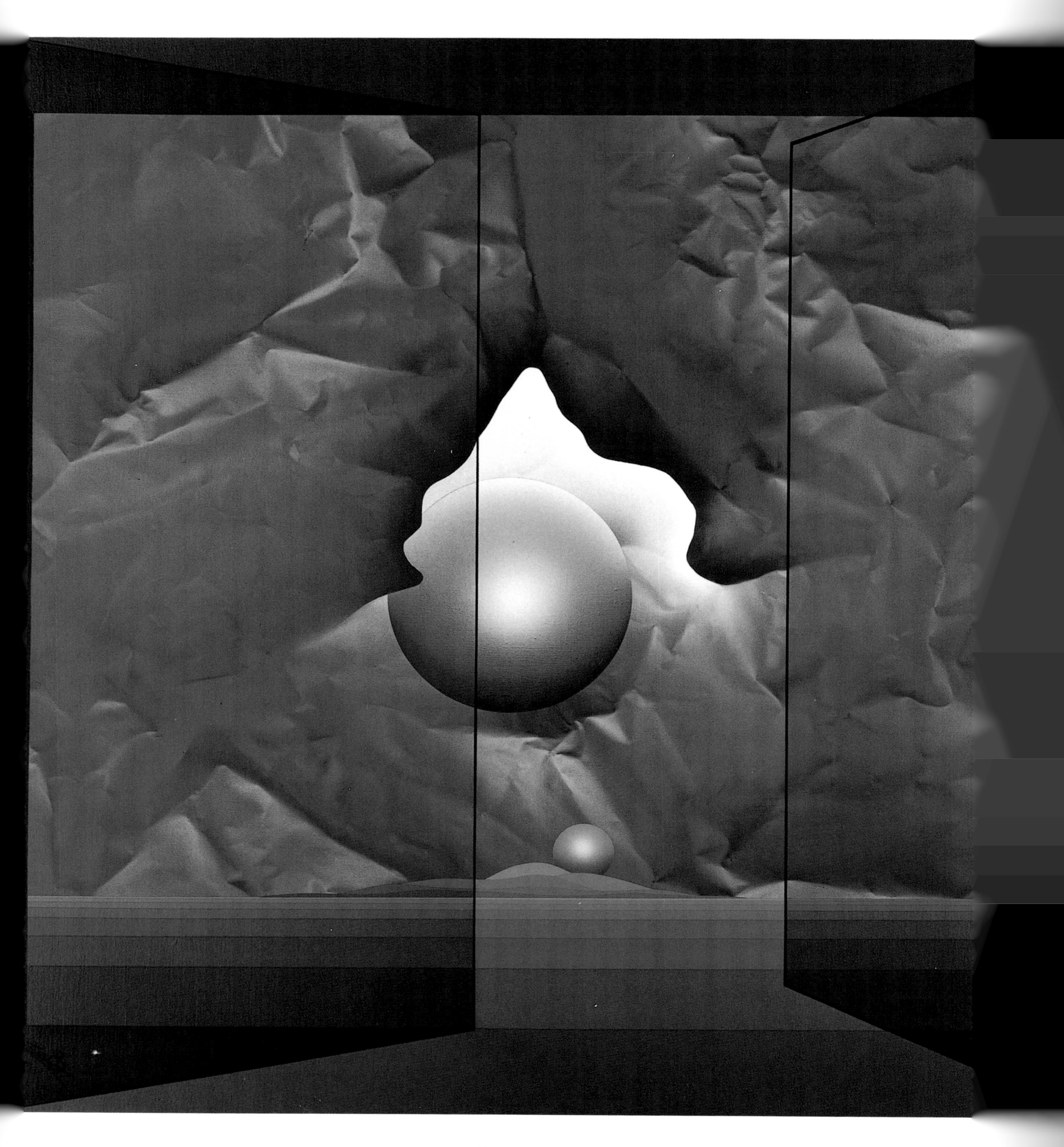

Doorway to Space
PATRICE BRETEAU

Not only had the satellite been a successful launch, it was also responsible for an important discovery. Explorer 1 carried aboard it a small instrument built by James A. Van Allen, a physicist from the University of Iowa. Essentially a Geiger counter, it measured radiation encircling the Earth and collected data that verified the existence of its magnetic field, now known as the Van Allen Radiation Belts—phenomena that partially dictate the electrical charges in the atmosphere and the solar radiation that reaches earth.

On February 1, a press conference took place at the National Academy of Sciences, where von Braun, Van Allen, and JPL director William H. Pickering announced their success. The signature image that appeared in newspapers around the nation the next morning depicted three smiling men triumphantly holding above their heads a full-scale model of Explorer 1, the United States' first artificial satellite. Project Vanguard subsequently received additional funding to accelerate its activity during this period, and it resulted in the orbiting of Vanguard 1 on March 17, 1958.

THE BIRTH OF NASA

As a direct outgrowth of the Sputnik crisis in the winter of 1957–1958, the administration worked with congressional leaders to draft legislation creating a permanent federal agency dedicated to exploring space. They placed all nonmilitary efforts relative to space exploration under a strengthened and renamed National Advisory Committee for Aeronautics (NACA).

Established in 1915 to foster aviation progress in the United States, the NACA had also been moving into space-related areas of research and engineering during the 1950's, through the work of a Space Task Group under the leadership of Robert L. Gilruth. Its civilian character, its recognized excellence in technical activities, and its quiet, research-focused image all made it an attractive choice. It could fill the requirements of the constrained job Eisenhower envisioned without exacerbating Cold War tensions with the Soviet Union.

During the summer of 1958, Congress passed the National Aeronautics and Space Act and the president signed it into law on July 29, 1958. This legislation created the National Aeronautics and Space Administration (NASA), effective October 1, 1958, and charged it with the broad mission to "plan, direct, and conduct aeronautical and space activities." Accordingly, less than a year after the launch of Sputnik, NASA's task involved the development of a far-reaching space exploration program.

CHAPTER ONE

BUILDING A SPACE PROGRAM

HEPARD
Alan B Shepard Jr
Bruce Stevenson

On October 1, 1958, T. Keith Glennan opened the doors to the National Aeronautics and Space Administration (NASA). Before 170 employees gathered in the courtyard of the Dolley Madison House near the White House, this newly appointed administrator unveiled his priorities for the space agency, announcing the bold prospects for space exploration under consideration by the nation at the time. They included robotic missions to the Moon and the planets of the solar system, communications satellites, and human exploration beyond the atmosphere. Like the undersea and polar discoveries then taking place, space exploration would require the development of new technologies as well as new social and political constructs.

Glennan was fortunate to preside over a newly established NASA that held the confidence of the American people and possessed a broad mandate for action. Resources flowed to the agency during those years, and few questioned the validity of the initiatives to explore the cosmos. Glennan wrote in his diary that "Congress always wanted to give us more money. . . . Only a blundering fool could go up to the Hill and come back with a result detrimental to the agency."

In an effort to assure the long-term viability of space exploration for the furtherance of the entire nation, Glennan effectively merged the National Advisory Committee for Aeronautics (NACA) into NASA, consolidating eight thousand employees, three research centers, and two flight stations, as well as elements of Department of Defense space research and development, and disparate space projects under other branches of the federal government. By the time he left NASA in January 1961, Glennan had brought to it the Army's Jet Propulsion Laboratory; the space exploration elements of the Naval Research Laboratory (which built Vanguard); and the space technology efforts at the DOD's Advanced Research Projects Agency, which was engaged in a project to develop a million-pound-thrust, single-chamber rocket engine.

But the jewel in Glennan's remarkable crown was the Army Ballistic Missile Agency (ABMA) in Huntsville, Alabama, in mid-1960. This 4,500-person installation, presided over by one of the nation's foremost space advocates, Wernher von Braun, was the Army's centerpiece in an interservice struggle for involvement in the space mission. The Army resisted Glennan's overtures for eighteen months, but congressional criticism eventually forced it to relinquish the ABMA. This maneuver, in addition to previous ones, proved critical in allowing NASA to undertake large-scale space exploration initiatives. The scientists and engineers who came to NASA from these other organizations brought with them a strong sense of technical competence, a

OPPOSITE:
Alan B. Shepard, Jr.
BRUCE STEVENSON
Shepard became the first American in space on MAY 5, 1961

PAGES 28–29:
In the Begining Nothing Became Everything
(Detail)
PAUL HUDSON

commitment to collegial in-house research conducive to engineering innovation, and a decidedly apolitical perspective.

Before leaving the federal government at the conclusion of the Eisenhower administration, Glennan had assured NASA's future as the agency charged with execution of all space exploration assignments. He had made plain the administration's desire to keep NASA nonmilitary in character, and had positioned it so that it would be able to complete far-reaching space missions. Yet Glennan was never convinced that the United States should explore space on essentially a blank check, or "crash" basis, advocating instead a measured space program that was incremental and never exceeded more than one percent of the federal budget. (It would reach more than five percent during the Apollo years and the race to the Moon, then decline precipitously thereafter to below one percent in 1998.) Ironically, Glennan succeeded almost too well in establishing NASA as a viable agency by enabling it to accomplish the type of large-scale technological enterprise he eschewed, such as an accelerated Project Apollo to race the Soviets to the Moon.

BEGINNING PROJECT MERCURY

By the spring of 1958, work had begun on the presidential assignment of putting a man in space. A group of NACA engineers, under the leadership of Dr. Robert Gilruth at the Langley Aeronautical Laboratory, proposed to Administrator Glennan a program designed to ascertain whether human beings could survive in the hostile space environment. With Glennan's subsequent approval, Project Mercury became the first major undertaking by the new agency.

During the months that followed, the Space Task Group aggressively pursued the development of the hardware and support structure necessary to handle the program. The Mercury spacecraft itself was genius incarnate, the brainchild of a diminutive Cajun named Dr. Maxime A. Faget, an engineering graduate of Louisiana State University and former submarine officer during World War II. Working at the NACA's Langley Research Center in Hampton, Virginia, Faget was one of the agency's most innovative and thoughtful engineers. While everyone who contemplated space flight in the 1950's was obsessed with rocket planes such as the X-15 and the X-20 Dynasoar, of

X-15
STAN STOKES
The rocket powered X-15 reached an altitude of 67 miles and speed in excess of 4,500 miles per hour (six and a half times the speed of sound).

Mission Control
MAXINE MCCAFFREY
Director of Flight Operations Chris Kraft and Robert Gilruth monitor a mission.

which the present space shuttle is the most advanced representation, Faget instead regarded space as an entirely different environment that required a comparably different type of vehicle.

Instead of worrying about lift-to-drag ratios—the bread and butter of modern aerodynamics—Faget considered a different tack. What if the United States built a vehicle that could skip in and out of the atmosphere? What if it built a cone with a relatively flat bottom to slow it as the craft entered the atmosphere—like a rock skipping on a lake. The blunt bottom would superheat to around 3,500 degrees, but with proper protection it would safely slow the spacecraft for a parachute landing on Earth. Faget knew that the technology existed to build such a craft, whereas a rocketplane flying from Earth into space and back—maneuverable in both flight regimes and at all speeds—was an enormous technological stretch in terms of materials and electronics. Equally important, any rocketplane would weigh more than the booster rockets envisioned in that early era.

Faget put six young engineers at Langley to work on a design for a conical space capsule capable of carrying a human above the atmosphere. They quickly developed a basic design, then spent several months refining it into the Mercury spacecraft. By no means elegant, this vehicle was a far cry from the science-fiction imaginings of the time, or even the proposed rocketplanes envisioned by such figures as Wernher von Braun.

Nevertheless, Faget showed his idea to his superior, Gilruth. As a remarkably successful aerodynamist who had headed the Pilotless Aircraft Research Division after World War II, Gilruth grasped the genius of this simple design and together the two obtained approval to proceed with it in October 1958.

On November 7, 1958, Faget, now chief designer of the Mercury spacecraft, held a briefing for forty aerospace firms to explain the requirements for bidding on the contract to build the capsule. Of twenty interested firms, eleven responded with proposals. With record-breaking speed almost never associated with phases of government procurement, Faget's team completed its evaluation after the Christmas holidays. The Source Evaluation Board recommended the McDonnell Aircraft Corporation of St. Louis, Missouri, as the prime manufacturer for this system. Glennan accepted this decision and announced the contract award on January 9, 1959.

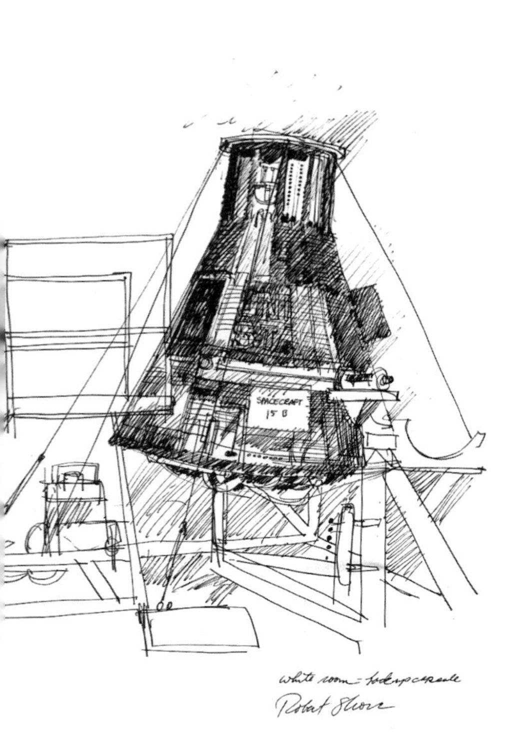

Backup Capsule in White Room
Robert Shore

Under the leadership of John F. Yardley, McDonnell's Mercury team immediately began wrestling with Faget's requirements. Although the Air Force had provided the team with a prior year's worth of good work, Yardley was unprepared for the difficulties surrounding the actual building of the spacecraft. First and most important, his team struggled with strict weight requirements for launching the capsule from atop the Atlas rocket. NASA's specifications called for placing a 2,700-pound capsule in orbit. McDonnell's bid had proposed a 2,400-pound spacecraft, plus or minus twenty percent. The minus side allowed for a capsule of 1,800 pounds, perfect for the capability of the Atlas, but anything over that could not be put into orbit by the envisioned launcher. A combination of paring the capsule design down to the lightest weight possible and increasing the thrust of the Atlas finally made successful Project Mercury launches attainable, but it was nevertheless a difficult task and the capability margins were always stretched.

There was also the problem of Atlas launch reliability. A converted ICBM, the Atlas had been undergoing an on-again, off-again development since 1946, having been canceled once and underfunded thereafter. It was not until the Sputnik crisis that the Air Force was able to secure sufficient resources to make serious progress on the missile. Because of this difficulty, the Atlas designers at the Convair Corporation accepted, as a given, a twenty percent failure rate. The rate, however, actually proved much higher in the early going. Seven out of eight launches in the beginning of 1959 failed, with some missiles blowing up on the pad or veering off flight course and having to be destroyed by the range safety officer. Instead of eighty percent reliability, which was still not acceptable for human flight, the Atlas had an eighty percent failure rate. A few months following the creation of the Space Task Group, Robert Gilruth testified before Congress about the failure-rate problem and urged legislators to provide the necessary time and money to test hardware under actual flight conditions without people aboard.

Ever so incrementally, Atlas project engineers improved the performance of the launch vehicle. They placed a fiberglass shield around the liquid oxygen tank to keep the engines from igniting it in a massive explosion, a rather spectacular failure that seemed to happen at least half the time. They changed virtually every system on the vehicle, substituting tried-and-true technology wherever possible to minimize problems. They altered procedures and developed new telemetry to monitor systems operation. Most important, they developed an abort-sensing system (labeled ASS by everyone other than

Liquid Fuel
PAUL SAMPLE
A gantry, with fuel tank in the foreground, is lit up in preparation for launch.

PAGES 36–37:
Suit Up
NORMAN ROCKWELL
John Young and Gus Grissom suit up for their Gemini 3 mission.

its developers) to monitor vehicle performance and to provide early escape from the Mercury capsule if necessary.

Simultaneously, Gilruth's engineers began integrating the Redstone and Atlas boosters with the spacecraft for practice flights. Only through this process would they determine whether they could operate reliably together. The first Mercury test flight took place on August 1, 1959, when a capsule carrying two rhesus monkeys was launched atop a cluster of Little Joe solid-fuel rockets. Other tests using both Redstone and Atlas boosters and carrying both chimpanzees and astronaut dummies soon followed. On January 31, 1961, Ham, a four-year-old chimpanzee, flew 157 miles into space in a 16-minute, 39-second flight in a Mercury-Redstone combination, and was successfully recovered.

THE ASTRONAUTS

During the massive technical effort taking place with the testing of Project Mercury components, NASA was also in the midst of selecting and training the Mercury astronaut corps. President Dwight D. Eisenhower directed that the astronauts be selected from among the armed services' test-pilot force. Although this was not NASA leadership's first choice, it greatly simplified the

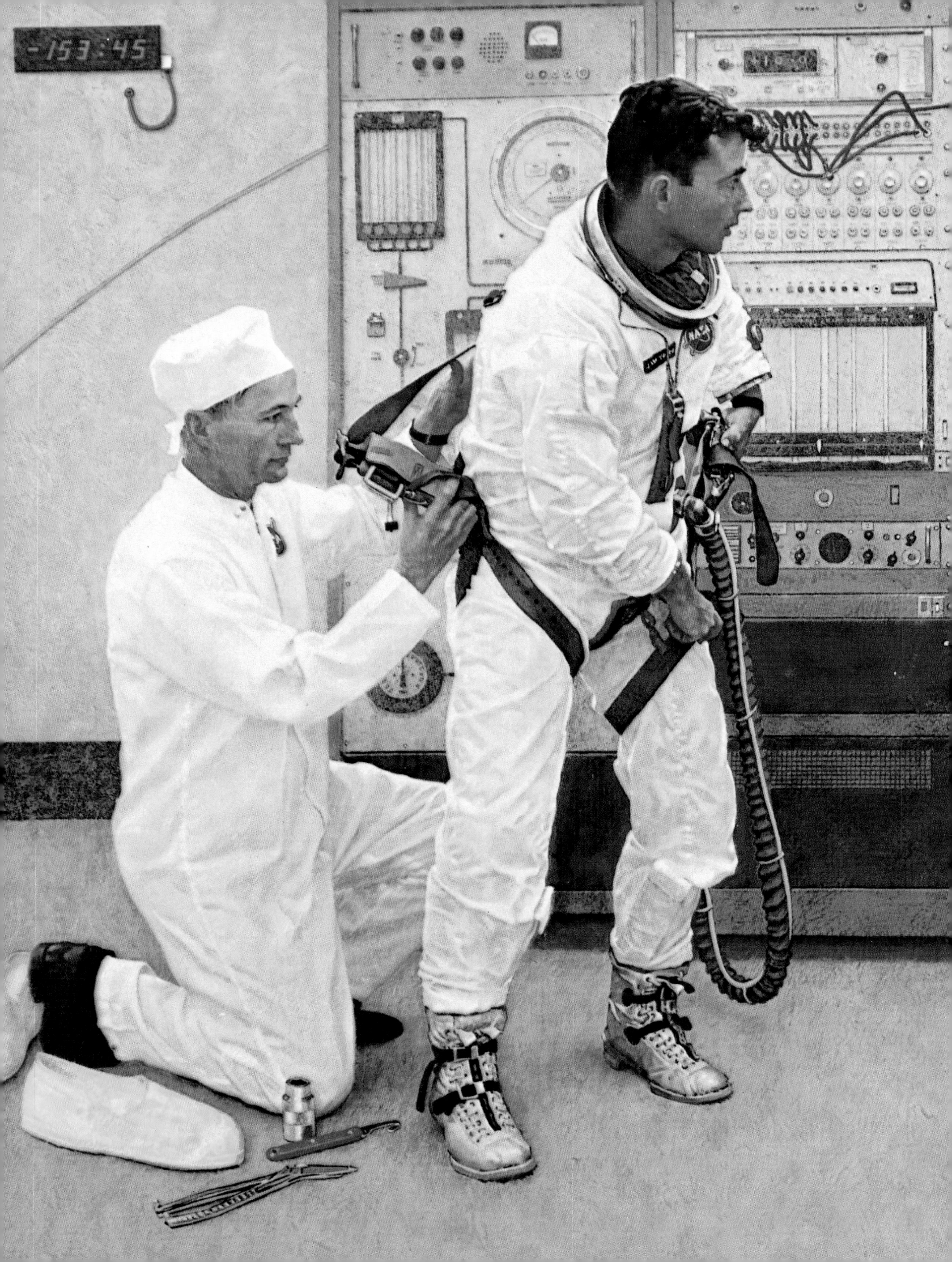
-153:45

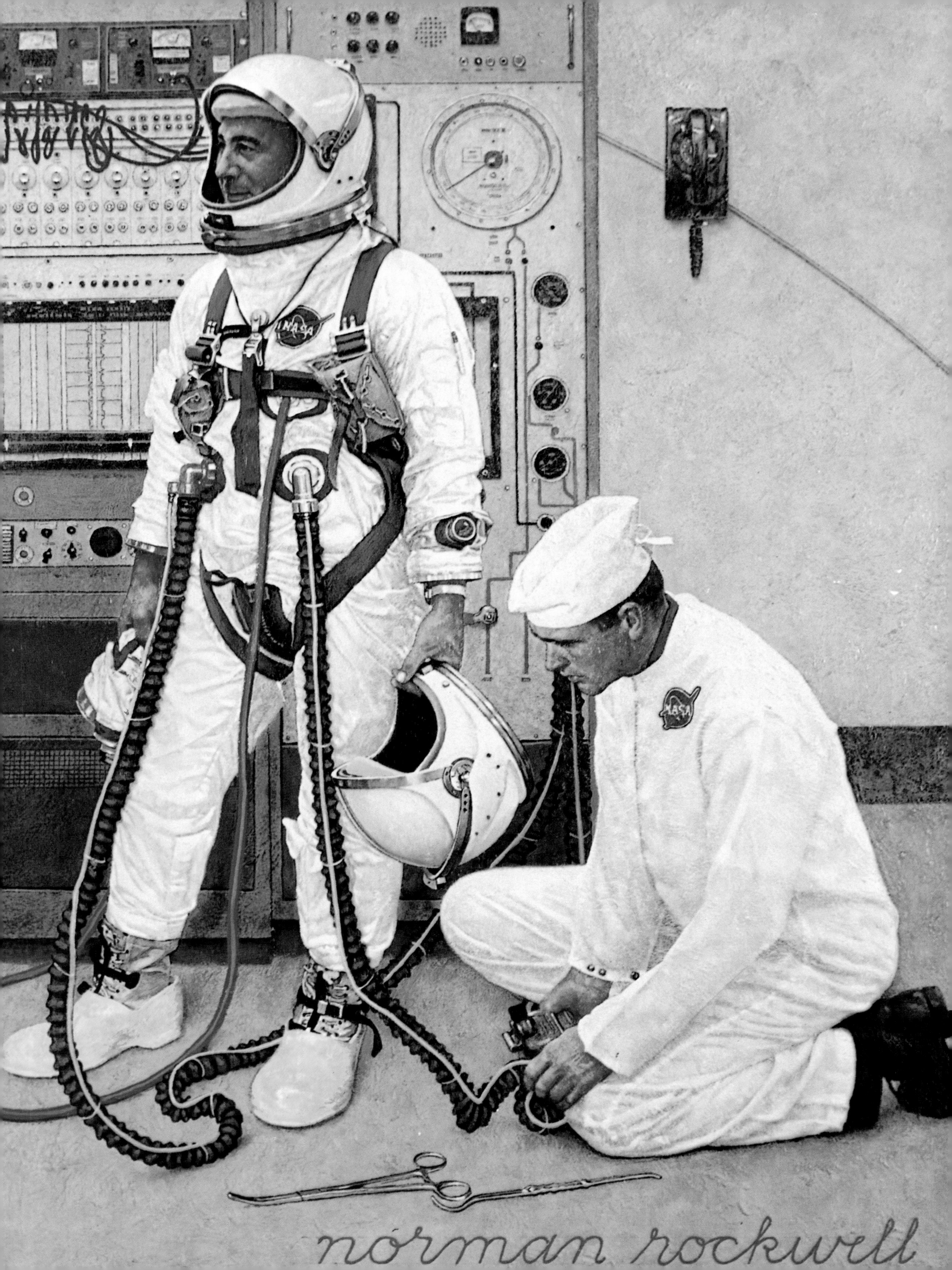
NASA
NASA
norman rockwell

selection procedure. During January 1959 NASA screened 508 service records at the military personnel bureaus in Washington from which 110 men—5 Marines, 47 Navy men, and 58 Air Force pilots—were found to meet the following established minimum standards for Mercury:

Age: under 40
Height: under 5'11"
Excellent physical condition
Bachelor's degree or equivalent
Graduate of test-pilot school
1,500 hours total flying time
Qualified jet pilot

Since the selection process began under the assumption that piloted Mercury-Redstone flights could take place late in 1959 (this later proved impossible), time was a critical factor. Headed by the Assistant Director of the Space Task Group, Charles J. Donlan, the evaluation committee divided the list of 110 into three groups and invited the first group of thirty-five to come to Washington in the beginning of February for briefings and interviews. Donlan's team initially planned to select twelve astronauts, but team member George M. Low believed in a lower number since few candidates would be expected to drop out, leaving many who would never ultimately participate in the flight program. It was agreed, therefore, to limit the finalists to six.

Each of the first ten pilots selected in February for interviewing agreed to continue through the elimination process. The next week the second group of candidates arrived in Washington. The high rate of volunteering made it unnecessary to invite the third group. Pilots selected from the first round of interviews were given a battery of written tests, technical and psychiatric interviews, and medical history reviews. By March 1, the number of candidates was reduced to thirty-six pilots.

The third phase of the selection process took place at the Lovelace Clinic in Albuquerque, New Mexico, and involved extraordinary physical examinations. Thirty different laboratory tests were conducted to collect chemical, encephalographic, and cardiographic data. X-ray examinations thoroughly mapped each body. The ophthalmology and otolaryngology sections were likewise extensive. Special physiological examinations included bicycle ergometer tests, total-body radiation count, total-body water determination, and the specific gravity of the entire body. Heart specialists performed complete cardiological examinations and other clinicians gathered perhaps the most extensive medical histories ever collected on human beings. After being certified as physically qualified by the Lovelace Clinic, thirty-two men accepted candidacy, knowing full well they were also scheduled for extreme mental and physical environmental tests at the Wright Air Development Center in Dayton, Ohio. (Remarkably, only one candidate had a medical problem that eliminated him from consideration.)

Phase four of the selection program involved an amazingly elaborate set of environmental studies, physical endurance tests, and psychiatric studies conducted at the Aeromedical Laboratory of the WADC. During March 1959 each of the candidates underwent a week of pressure-suit, acceleration, vibration, heat, and loud-noise tests. The Dayton experience involved continuous psychiatric interviews, the necessity of living with two psychologists throughout the week, an extensive examination through a battery of thirteen personality and motivation tests, and another dozen tests on intellectual functions and special aptitudes.

Two of the more interesting personality and motivation studies seemed like parlor games at first, until the pilots realized the importance of their answers to the questions "Who am I?" and "Whom would you assign to the mission if you could not go yourself?" The first question required the subject to write down twenty identifications of himself, ranked in order of significance. Interpreted productively, the results provided information on self identity and perception of social roles. A set of questions investigating peer rating included asking the candidate whom of his fellow candidates he liked best and which he would like to accompany on a two-man mission.

Despite their thoroughness, these tests did not yield conclusive results. In late March 1959, Gilruth's Space Task Group began phase five of the selection, narrowing the candidates to eighteen. The final selection criteria thereafter reverted to the men's technical qualifications and the requirements of the program as judged by Donlan and his team. "We looked for real men and valuable experience," said Donlan. The verbal responses made at the interviews, therefore, seem to have been as important final determinants as the candidates' test scores.

Donlan's evaluation team finally narrowed the eighteen finalists to seven, but they agonized over the last necessary cut to six. Since they could not reach that magic number, Gilruth decided to recommend seven. Donlan telephoned each of the finalists individually to reconfirm his willingness to become a Mercury astronaut. All gladly volunteered again. Unable himself to cut the last candidate, Glennan agreed to appoint seven astronauts. Gilruth invited those not chosen to reapply for reconsideration in some future program. Many ultimately did, including the first human on the Moon, Neil A. Armstrong.

THE MERCURY PRESS CONFERENCE

Almost immediately, and despite the wishes of NASA leadership, the seven chosen men became heroes in the eyes of the American public. Their fame quickly grew beyond all proportion to their assignments—to most Americans, they *were* NASA, in no small part because of the deal they made with *LIFE* magazine, granting it exclusive rights to their stories.

Perhaps this premature adulation was inevitable when one considers the enormous public curiosity about the astronauts, the great risks they would take in attempting space flight, and the exotic training activities they underwent. But the power of commercial competition for publicity and the pressure for political prestige in the space race also whetted an insatiable public appetite for this new kind of celebrity. Walter Bonney, long a public information officer for NACA and now Glennan's adviser on these matters, foresaw the public and press attention and asked for an enlarged staff. He then proceeded to lay down guidelines for a public-affairs policy that was in close accord with those of other government agencies.

Bonney's foresight proved valuable on April 9, 1959, when NASA chose to unveil the first Americans to fly in space. Excitement bristled in

Press Conference at Cape Canaveral, May 19, 1963
MITCHELL JAMIESON

Washington at the prospect of learning who those space travelers might be. Surely they were the best the nation had to offer, modern versions of medieval knights whose honor and virtue were beyond reproach. They carried on their shoulders all the hopes, dreams, and best wishes of a nation.

NASA's makeshift Washington headquarters was abuzz with employees hastily setting up a temporary press-briefing room out of what was once a ballroom on the second floor. One end of the room sported a stage complete with curtain, and both NASA officials and the newly chosen astronauts waited behind it for the press conference to begin at 2:00 P.M. The other end had trip hazards of electrical cable strewn about the floor, banks of hot lights mounted to illuminate the stage, and more than a few television cameras carrying the event live and movie cameras recording footage for later use. Members of the media were jammed in their cramped area, with photographers gathered at the foot of the stage and journalists occupying seats in the gallery.

When the curtains went up, NASA public affairs officer par excellence, Walter Bonney, announced, "Ladies and gentlemen, may I have your attention, please. The rules of this briefing are very simple. In about sixty seconds we will give you the announcement that you have been waiting for: the names of the seven volunteers who will become the Mercury astronaut team." Immediately, photographers moved forward and popped flashbulbs in the faces of the astronauts. The buzz in the conference room rose to a roar as the photo shoot proceeded. Some of the journalists bolted for the door with their press kits to file their stories for the evening papers; others ogled the astronauts.

Fifteen minutes later Bonney brought the room to order and asked Keith Glennan to come out and formally introduce the astronauts. Glennan offered a brief welcome and added, "It is my pleasure to introduce to you—and I consider it a very real honor, gentlemen—Malcolm S. Carpenter, Leroy G. Cooper, John H. Glenn, Jr., Virgil I. Grissom, Walter M. Schirra, Jr., Alan B. Shepard, Jr., and Donald K. Slayton . . . the nation's Mercury astronauts!" These personable pilots were introduced in civilian dress; many people in their audience forgot that they were volunteer test subjects and military officers. Rather, they were a contingent of adult middle-class Americans, average in build and visage, all family men, college-educated as engineers, possessing excellent health, and professionally committed to flying advanced aircraft.

The reaction was nothing short of an eruption. Applause drowned out the rest of Glennan's remarks. Journalists rose to their feet in a standing ovation. Even the photographers crouched at the foot of the stage rose in acclamation of the Mercury Seven. At first, the newly selected astronauts replied to the press corps' questions with military stiffness, but led by an effervescent and sentimental John Glenn, they soon warmed to the interviews.

What seemed to surprise the astronauts most was the nature of the questions most often asked. Although all the men had been military test pilots—many with combat experience, decorations for valor, and even aircraft speed and endurance records—reporters did not seem to care about flying experience. Few wanted the details of NASA's plans for Project Mercury. What did interest them greatly were the astronauts' personal lives. The media wanted to know about their devotion to the country; whether they believed in God and practiced any religion; if they were married and the names and ages and gender of their children; and what their families thought about space exploration and the astronauts' roles in it. God, country, family, and self—reporters searched for confirmation that these seven men embodied the deepest virtues of the United States. They wanted to demonstrate to their readers that the Mercury Seven strode the Earth as latter-day saviors whose purity coupled with noble deeds would purge this land of the evils of communism by besting the Soviet Union on the world stage.

John Glenn, perhaps intuitively or perhaps through sheer zest and innocence, picked up on the mood of the audience and delivered a ringing sermon on God, country, and family that sent the reporters rushing to their phones for rewrite. "I think we would be most remiss in our duty," he concluded, "if we didn't make the fullest use of our talents in volunteering for something that is as important as this is to our country and to the world in general right now." The other astronauts fell in behind Glenn and eloquently spoke of their sense of duty and destiny as the first Americans to fly in space. Near the end of the meeting, a reporter asked if they believed they would come back safely from space and all raised their hands. Glenn raised both of his.

By the next morning the Mercury Seven were household names and heroes to virtually every schoolchild in America. The reporters had cast the

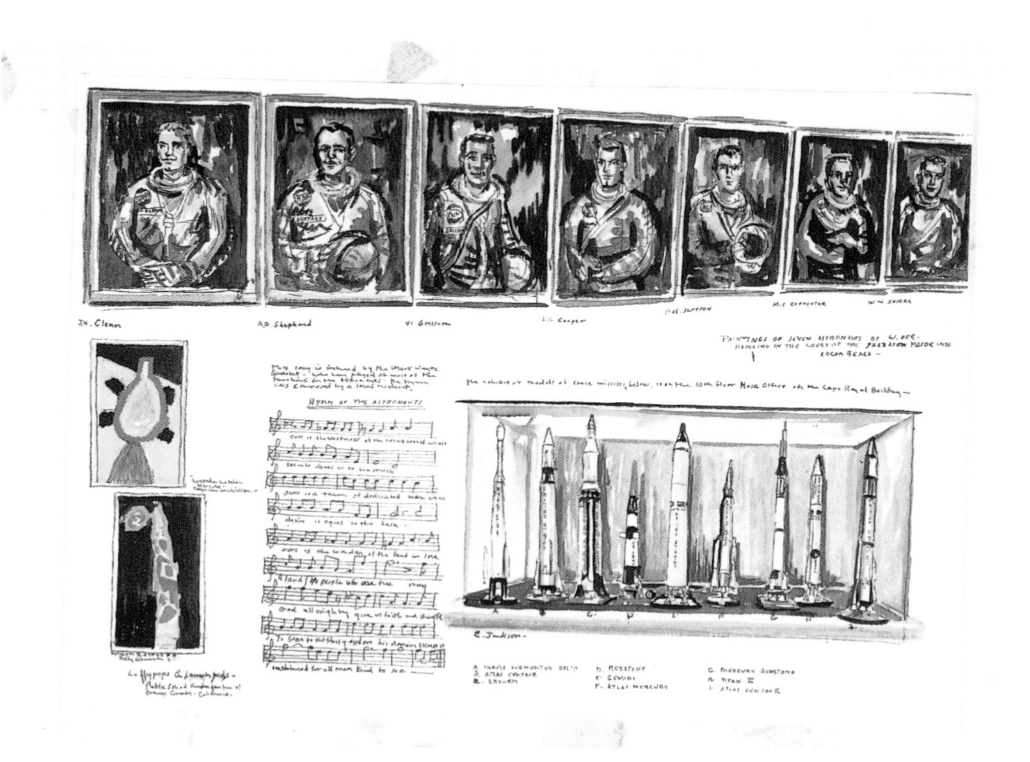

Cape Canaveral Scenes
Crystal Jackson
This tribute to the Mercury Seven includes portraits originally painted by W. Orr. Appearing left to right: John H. Glenn, Alan B. Shepard, Virgil I. Grissom, Leroy G. Cooper, Donald K. Slayton, Malcolm S. Carpenter, and Walter M. Schirra.

OPPOSITE:
John Glenn
Bruce Stevenson
John Glenn became the first American to circle the Earth, making three orbits in the Mercury spacecraft Friendship 7.

astronauts as champions who would carry the nation's manifest destiny beyond its shores and into space. James Reston of the *New York Times*—a newspaper with a history of criticizing space flight all the way back to Robert Goddard—exulted over the astronaut team. "What made them so exciting," he wrote, "was not that they said anything new but that they said all the old things with such fierce convictions. . . . They spoke of 'duty' and 'faith' and 'country' like Walt Whitman's pioneers . . . This is a pretty cynical town, but nobody went away from these young men scoffing at their courage and idealism."

The Mercury Seven were, in essence, each of us. None was aristocratic in bearing or elitist in sentiment. They came from throughout the nation, excelled in the public schools, trained at their local state university, served their country in war and peace, married and tried to make lives for themselves and their families, and ultimately rose to their places on the basis of merit. They represented the best we had to offer, and, most important, they expressed at every opportunity the virtues ensconced in the democratic principles of the republic.

ACCOMPLISHING MERCURY

With the spacecraft and rocket on their way to completion, and the astronauts selected and now in flight training, NASA needed to maintain a sensible momentum toward the goal of space capability. Unfortunately, the public—and many in government—were more interested in spectacular developments. In fact, late in 1959, Glennan incurred the wrath of congressional members when it appeared that the Mercury Program was not moving fast enough. The joke in Washington at the time was that the first man in space would be neither a Soviet cosmonaut nor an American astronaut but Glennan, who would be launched by Congress unless he got NASA moving more quickly. Glennan instead tried to persuade politicians to build a broad-based program that would yield valuable scientific and technological information rather than achieve an impressive but less substantive result.

Glennan did not want to conduct U.S. space program activities in response to the Soviet Union. His diary reads, "We are not going to attempt to compete with the Russians on a shot-for-shot basis in attempts to achieve space spectaculars. Our strategy must be to develop a program on our own terms which is designed to allow us to progress sensibly toward the goal of ultimate leadership in this competition." He even proposed that the United States enter a cooperative space research project with the Soviet Union, but the Cold War environment of the time made the idea impossible to carry out.

To establish a clear direction for NASA, Glennan's staff developed a ten-year plan emphasizing the scientific and technological developments to be attained in each of the following areas: space-vehicle development, manned space flight, engineering and scientific research, and space-flight operations. The program called for approximately $12.5 billion in 1959 dollars to accomplish a hefty scientific probe program, a human space-flight program that would launch its first astronaut in 1961, and a lunar landing at some unspecified time in the post-1970 period. It also provided for the development of new launch boosters that would give the United States a decided edge in long-term space activities.

This was a modest but reasonable and optimistic program, and the Eisenhower administration accepted the funding priorities. Glennan wrote in his diary, "Ike and I agreed that we were mature enough as a nation not to let some other country determine our behavior and policy. Hence, we opposed a 'Space Race,' and while we wanted to advance rapidly, not to do foolish things just because the Russians were doing them." At the same time, Glennan recognized that the United States could not operate in a business-as-usual mode with the Soviets influencing world opinion in their favor by executing "space spectaculars." He opted, instead, for a deliberate program with clear objectives and a long timetable.

A centerpiece of this NASA space exploration effort was the successful completion of Project Mercury. By the spring of 1961, Gilruth and his Space Task Group confidently predicted that they stood at the dawn of human space flight. The astronauts had prepared as well as they could for the rigors of the harsh new environment. Despite difficulties, the Mercury spacecraft finally appeared ready to go. Even the troubled Atlas launcher program seemed successful enough to allow Gilruth to exude optimism. By May, at the latest, the first American astronaut would fly the first suborbital mission.

A deliberate pace might have remained the standard for Project Mercury had not the Soviet Union's space effort scored a coup on April 12, 1961, when Soviet cosmonaut Yuri Gagarin became the first human in space with a one-orbit mission aboard the spacecraft Vostok 1. The chance to place a human in space before the Soviets had been lost.

NASA's Space Task Group redoubled its efforts to launch the first Mercury flight after Gagarin's success. At 9:34 A.M. on May 5, 1961, nearly forty-five million Americans sat tensely before their television screens and watched as Freedom 7—a slim black-and-white Redstone booster, capped with a Mercury spacecraft containing astronaut Alan Shepard—lifted off its pad at Cape Canaveral and roared upward through the blue sky. At 2.3 sec-

Cosmonaut Yuri Gagarin
PAUL CALLE
Star City, Moscow.
Soviet cosmonaut Yuri Gagarin, the first human in space.

onds after launch, Shepard's voice came through clearly to Mercury Control; minutes later the millions heard the historic transmission, "Ahh, Roger; liftoff and the clock is started . . . Yes, sir, reading you loud and clear. This is Freedom 7. The fuel is go; 1.2 g; cabin at 14 psi; oxygen is go . . . Freedom 7 is still go!"

Shepard, the first American in space, was in flight for 15 minutes and 22 seconds and was weightless only a third of that time. Freedom 7 rose to an altitude of 116.5 miles, attained a maximum speed of 5,180 miles per hour, and landed 302 miles downrange from the Cape. The success of this flight, however, was but a small consolation to many Americans and citizens of other nations who realized or celebrated the accomplishment of the communist state with its world-leading technology.

On July 21, 1961, a second suborbital Mercury flight, the Liberty Bell 7, was launched on July 21, 1961. Unfortunately, it proved less successful than Shepard's mission. After landing, the hatch blew off prematurely from the capsule, causing it to sink into the Atlantic Ocean before it could be recovered. Astronaut "Gus" Grissom nearly drowned before a helicopter hoisted him to safety. Despite success or failure, however, these suborbital flights proved valuable for NASA technicians who found ways to work through thousands of obstacles to successful space flight.

As NASA engineers were resolving the capsule's technical problems, they also began final preparations for the orbital aspects of Project Mercury. For this phase, NASA planned to employ a Mercury capsule that could support a

MISSION AND LAUNCH CONTROL

During the course of its thirty-five years of manned space flight, the meters, knobs, and black-and-white displays of the original Mission Control Center (MCC) have given way to electronic documentation, PC mouses, color graphics, and artificial intelligence. These changes, however, have never altered Mission Control's awesome task: to control and direct astronauts and spacecraft with authority, responsibility, and safety.

Fulfilling this directive requires the cooperation and coordination of hundreds of personnel. Each key system of a spacecraft—propulsion, guidance, and mechanical systems, electrical power, thermal/environment, communications, and others—is assigned to a flight controller who is responsible for its monitoring, operation, and safety, as well as ensuring that the respective system is operating in accord with other systems.

The figure at the head of this massive effort is the Flight Director, who receives the recommendations and technical judgments from the flight controllers and makes the ultimate decisions during a mission. Any course of action is relayed by the Flight Director to the CAPCOM, or "capsule communicator." The CAPCOM, the only astronaut on the flight-control team, then voices the instructions to the crew.

From liftoff to landing, MCC operates around the clock by utilizing three flight-control teams, each headed by its own flight director and operating overlapping nine- to ten-hour shifts. Twelve hours before launch, mission-control teams are at the consoles and monitoring the countdown, while Kennedy Space Center Launch Control Center (LCC) goes through its launch operations.

Once the shuttle engines ignite, authority for the mission shifts to MCC, which is then in charge of the ascent into orbit, all the on-orbit operations, reentry, and landing. Should a problem arise, telemetry information is examined and the various flight controllers report their findings and recommendations to the flight director who then directs the crew in working through the problem. Once the astronauts exit the shuttle following landing, Mission Control passes responsibility of the shuttle back to Kennedy Space Center.

During NASA's earlier manned programs, MCC communicated with the astronauts through a complex worldwide tracking network. Astronauts were able contact Mission Control only when their spacecraft was positioned above a tracking station. Since astronauts would be traveling 150 miles above the earth at 5 miles per second, communication passes typically lasted five minutes or less and occurred only several times each orbit, amounting to only fifteen to twenty percent duration of contact during a mission. Today, Tracking and Data Relay Satellites (TDRS)—positioned over the equator at 22,000 miles above the earth in geosynchronous orbit and monitored at White Sands, New Mexico—relay voice transmissions, telemetry data, commands, and tracking information between MCC and the Shuttle, allowing MCC contact with a spacecraft during ninety percent of a mission's duration.

Mission Control is by no means idle between launches. It takes about one and a half years to put a mission together, as controllers write flight procedures, mission rules, and timelines, develop flight-specific software, and run through coordinated simulations involving the astronauts and the flight team.

ROB KELSO, FLIGHT DIRECTOR, MISSION CONTROL CENTER

Firing Room
JOHN MEIGS

LEFT AND OPPOSITE:
Mission Control, Apollo 11
FRANKLIN MCMAHON

Firing Room
James Wyeth
Launch crew members carefully monitor the Apollo 11 countdown from Firing Room 1 at Kennedy Space Center.

Mercury-Atlas 9, May 1963
PAUL CALLE
Prelaunch activity continues into the night.

OPPOSITE:
Gantry Structure
LAMAR DODD

human in space not only for minutes but for as long as three days—an extended capability that would greatly enhance NASA's knowledge of the problems of space flight. To this end, the first successful orbital flight of an unoccupied Mercury-Atlas combination was launched in September 1961. On November 29 the final test flight took place with a chimpanzee, Enos, which resulted in a two-orbit ride and a successful ocean landing and recovery.

On February 20, 1962, NASA finally launched an orbital flight using an astronaut, and on that day, John Glenn became the first American to circle the Earth, making three orbits in his Friendship 7 Mercury spacecraft. The flight did have its problems, however. Due to autopilot failure, Glenn flew portions of the last two orbits manually. He was also forced to leave his normally jettisoned retrorocket pack attached to the capsule during reentry because of a loose heat shield.

Regardless, Glenn's flight increased national pride and made up for at least some of the Soviet's early successes. The public, more than celebrating technological success, embraced Glenn as the personification of heroism and dignity. Hundreds of requests for personal appearances poured into NASA

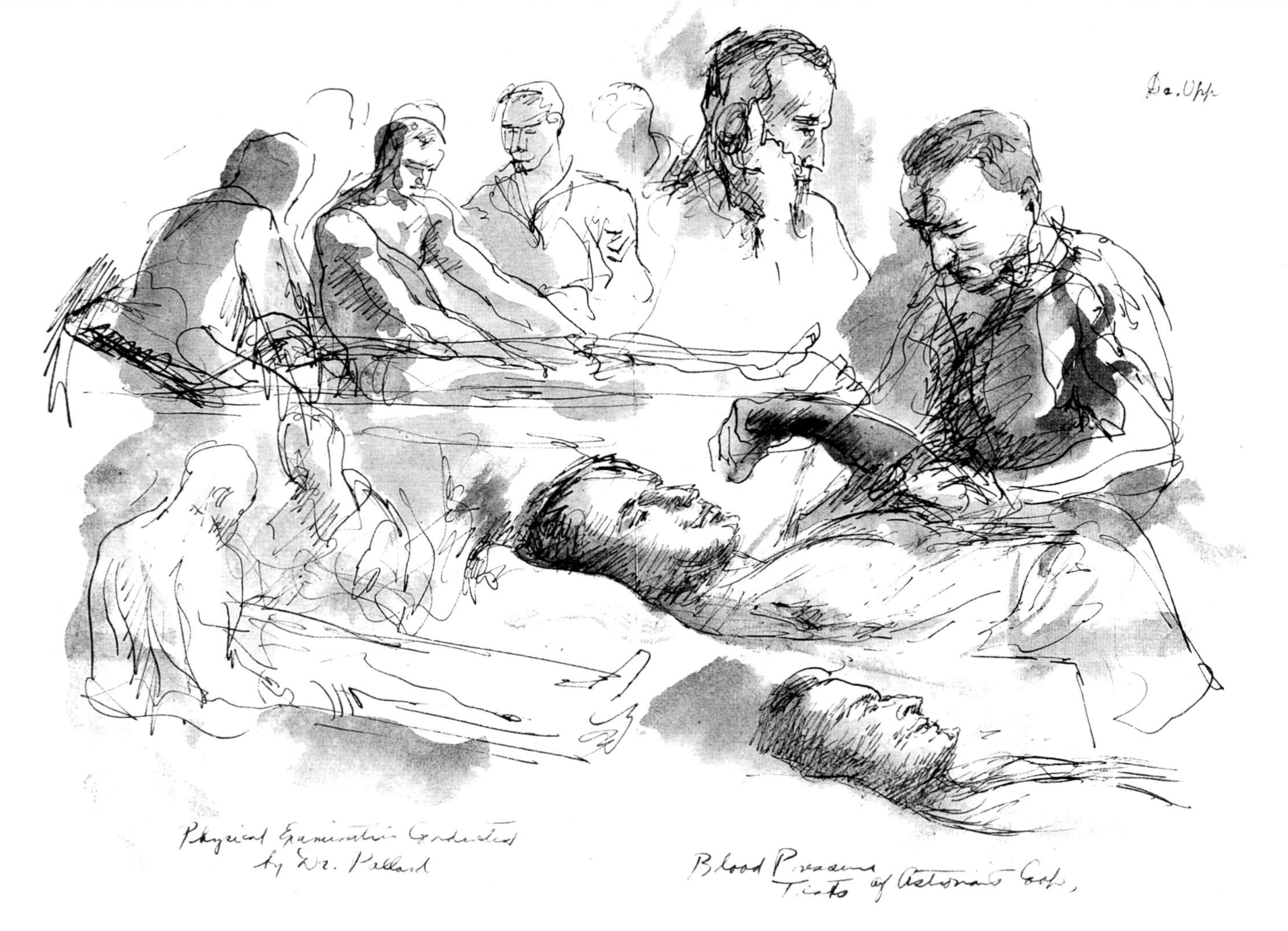

Medical Examination of Astronaut Cooper Aboard USS Kearsarge.
MITCHELL JAMIESON

RIGHT:
Pressure Suit Test at Hanger "S", Mercury-Atlas 9 Launch
PAUL CALLE

OPPOSITE:
First Steps
MITCHELL JAMIESON

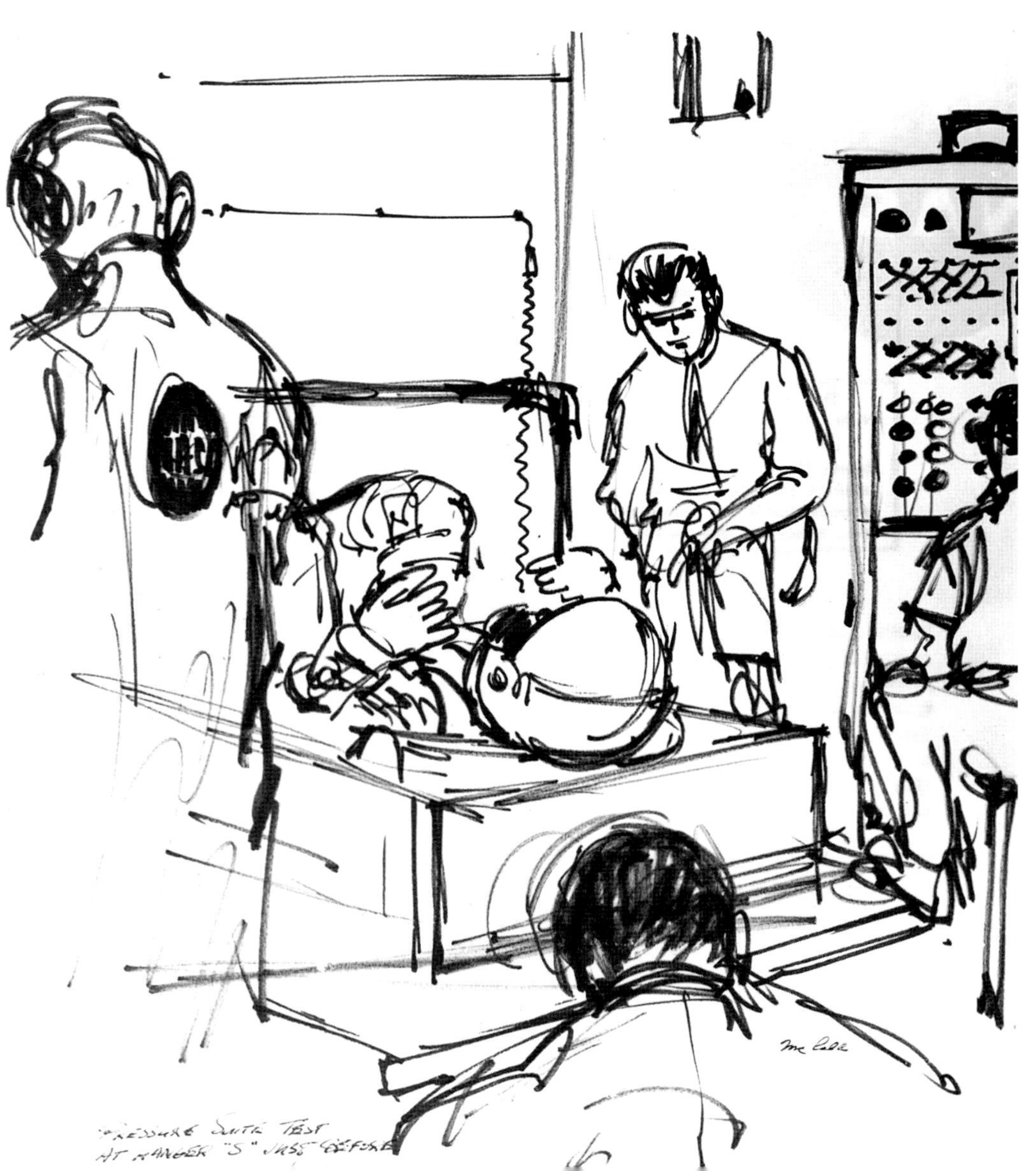

First Light
John W. McCoy

opposite:
Pre-Dawn
Peter Hurd

"Whether by design, by chance, or from technical need, the score or more of enormous gantry cranes, which seemed to stride in a great marching procession along the shore, were painted an intense and subtly beautiful shade of red. The cranes are of open steel work and interlacing maze of girders and tubing, lavishly lighted from inside and out, giving an unbelievably realistic effect of incandescent filigree."

Artist Peter Hurd

From the Inside
Robert Shore
A close look at the inner workings of a gantry structure.

"It was as if I stepped into the fantastic world of science fiction come alive. Jules Verne come true! "

Artist Robert Shore

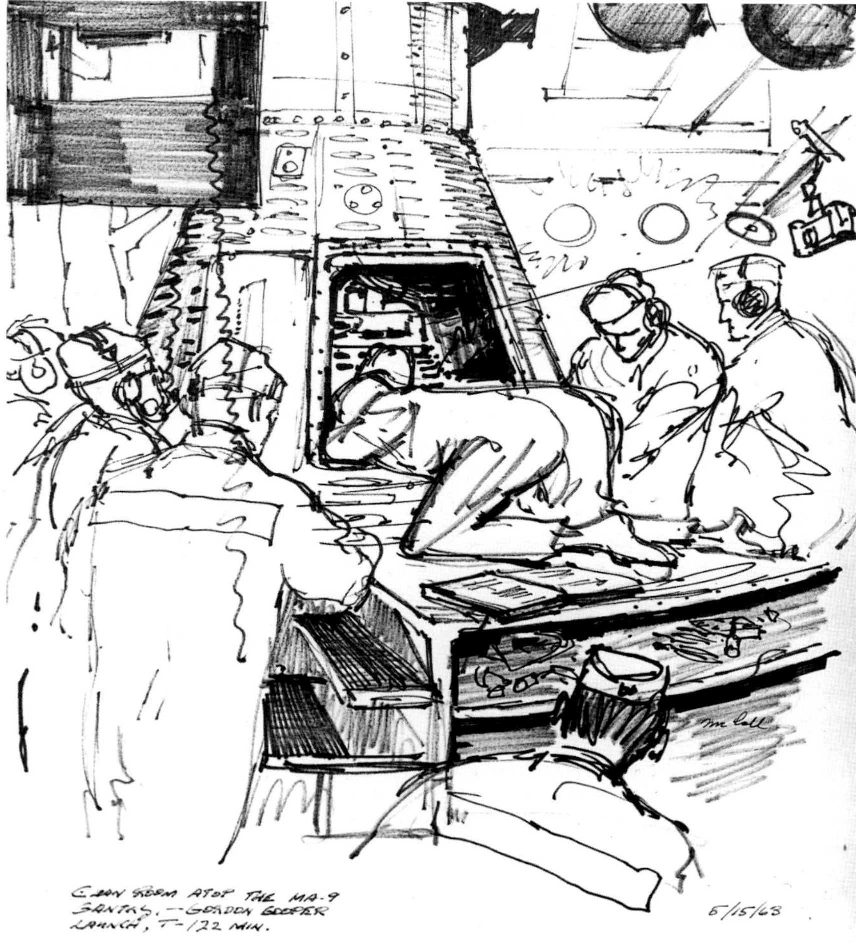

The Sounds of Engines
THEODORE HANCOCK
Rocket engine tests at Marshall Spaceflight Center in Huntsville, Alabama.

LEFT:
Mercury-Atlas 9 Spacecraft, Faith 7
ROBERT MCCALL
Preparing the spacecraft for launch.

Max Q
LAMAR DODD
A Mercury astronaut, strapped in and ready for launch.

OPPOSITE:
Mercury-Atlas 9
JOHN W. MCCOY
Mercury Faith 7, launched May 15, 1963.

headquarters and the agency learned just how influential astronauts could be in swaying public opinion. NASA leadership made Glenn available for some speaking engagements, but because of the large number of requests, other astronauts were also substituted. Among his appearances, Glenn also addressed a joint session of Congress and participated in several ticker-tape parades around the country.

Three more successful Mercury flights took place during 1962 and 1963, including Scott Carpenter's three-orbit completion on May 20, 1962, and Walter Schirra's six-orbit flight on October 3, 1962. The capstone of Project Mercury, however, was the May 15–16, 1963, flight by Gordon Cooper, who circled the earth twenty-two times in thirty-four hours.

Cooper's flight marked the successful end of Project Mercury and the accomplishment of its three simple goals: to orbit a human successfully in space, to explore aspects of tracking and control, and to learn about microgravity and other biomedical issues associated with space flight. Having run a

First Sighting
MITCHELL JAMIESON

Naval shipmen, stationed on a recovery ship in the Pacific Ocean, observe the descending Mercury spacecraft.

SPACECRAFT RECOVERY

The first task of any space vehicle with people aboard is to return them safely to Earth. That return was not always elegant, but for the U.S. space program it has always been successful. Unlike the Soviet Union, which lost three cosmonauts during a reentry to Earth in 1971, the first American missions—Mercury, Gemini, Apollo, and Skylab—all used a currently outmoded recovery method. Flying at some 16,000 miles per hour in Earth orbit, an astronaut would position the spacecraft so that its blunt body faced forward and fire a retrorocket pack that would slow the vehicle down and decay its orbit. As the craft entered the Earth's atmosphere, the air would create even greater drag. To protect the crew, a heat shield, consisting of ablative materials that burned away, allowed the spacecraft to remain intact. Once a sufficiently slow speed was achieved, a parachute would release from the nose of the spacecraft and the entire vehicle would land in the ocean. Navy ships tracking the spacecraft as it descended from orbit dispatched helicopters with frogmen to recover the spacecraft and its crew for a triumphal return to the United States.

This process was expensive and time-consuming. It would have been much better to land the space vehicle on a runway, as is currently done with the shuttle, but the method worked well and no astronauts were ever lost in the process. This was due, in part, to a well-proven technology. The blunt-body (essentially teardrop) shape of early space vehicles was developed during the early 1950's, when research for the reentry of nuclear warheads proved important to the ballistic missile programs. One disadvantage to this design, however, was that the outer heat shields reached temperatures in excess of 1,000 degrees upon reentry. The physics of this meant that a hot ionized flow engulfed the spacecraft with electrically charged particles that formed a "plasma sheath" behind the bow shock. This would result in a radio transmission "blackout" for several minutes during reentry, generating great suspense for ground controllers and interested observers waiting to learn the fate of the crew.

The way in which NASA accomplished reentry during early missions will probably not be repeated. "We were pretty lucky," one leader in Project Mercury recalled. "In retrospect, we wouldn't dare do it again under the same circumstances. But that's true of most pioneering ventures. You wouldn't dare fly across the ocean with one engine like Lindbergh did, either, would you?"

Roger Launius, NASA Chief Historian

Apollo 10 Boilerplate Is Lifted
CHET JEZIERSKI

Bringing in the Boilerplate After Simex
TOM O'HARA

OPPOSITE:
Frogmen Attach Lines to Apollo 9 Spacecraft
LEONARD DERMOTT

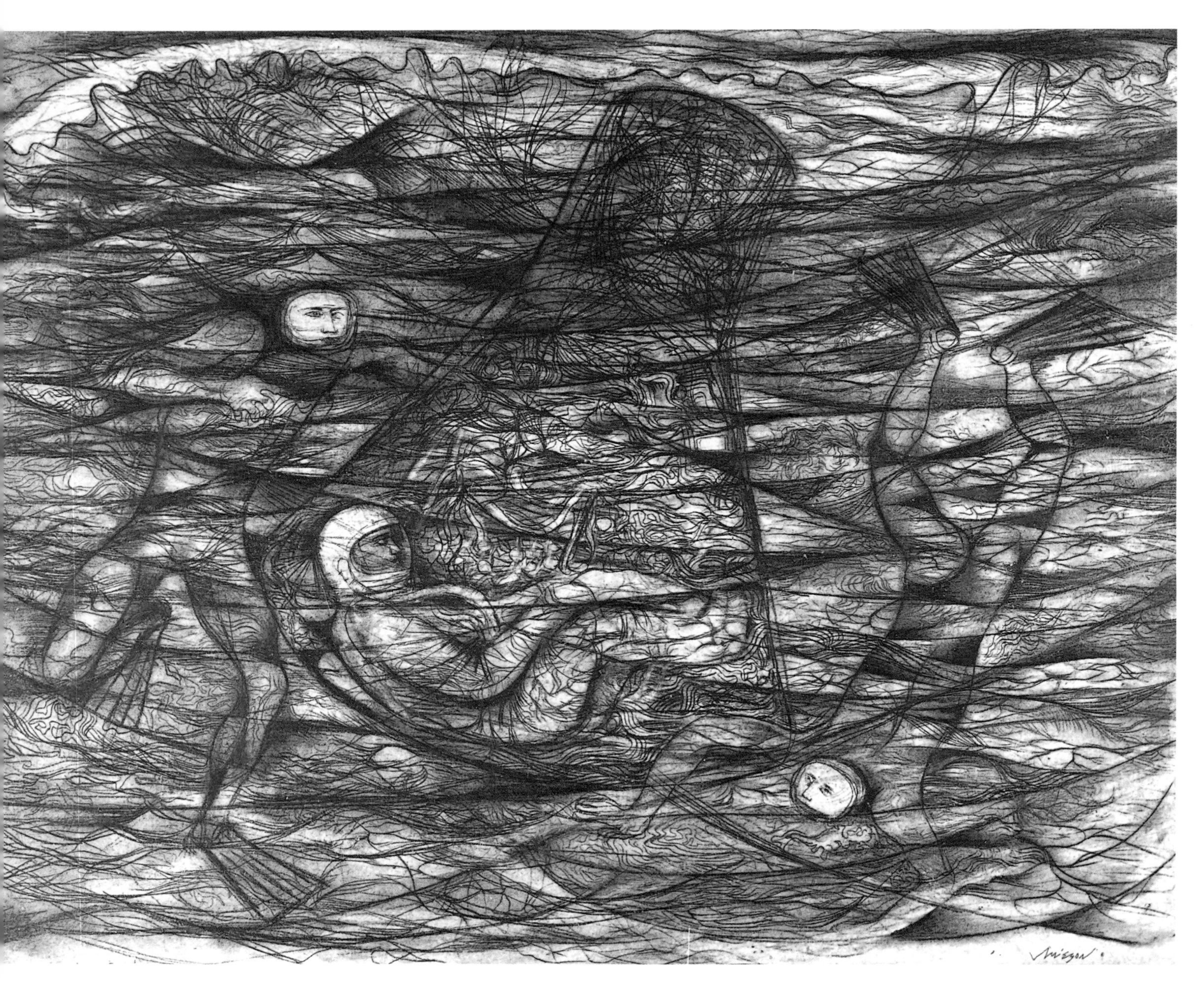

Pacific Recovery
MITCHELL JAMIESON

OPPOSITE:
Afterthoughts, L. Gordon Cooper
MITCHELL JAMIESON
Gorden Cooper during the final Project Mercury mission.

little less than five years at a cost of about $400 million, the program was a real bargain for human space flight. And although it lagged behind Bob Gilruth's original schedule, it succeeded in proving the possibility of safe human space exploration and in demonstrating to the world U.S. technological competence during the Cold War.

As one then-congressman concluded, Project Mercury reminded him of a "Rube Goldberg contraption placed on top of a plumber's nightmare." An accurate depiction to be sure, but as the Mercury Program progressed from 1961 to 1963, NASA's greatest challenge took shape. In May 1961, President John F. Kennedy's promise to the American people gave NASA the charter to accomplish Project Apollo's objective: land an American on the Moon before the end of the 1960's.

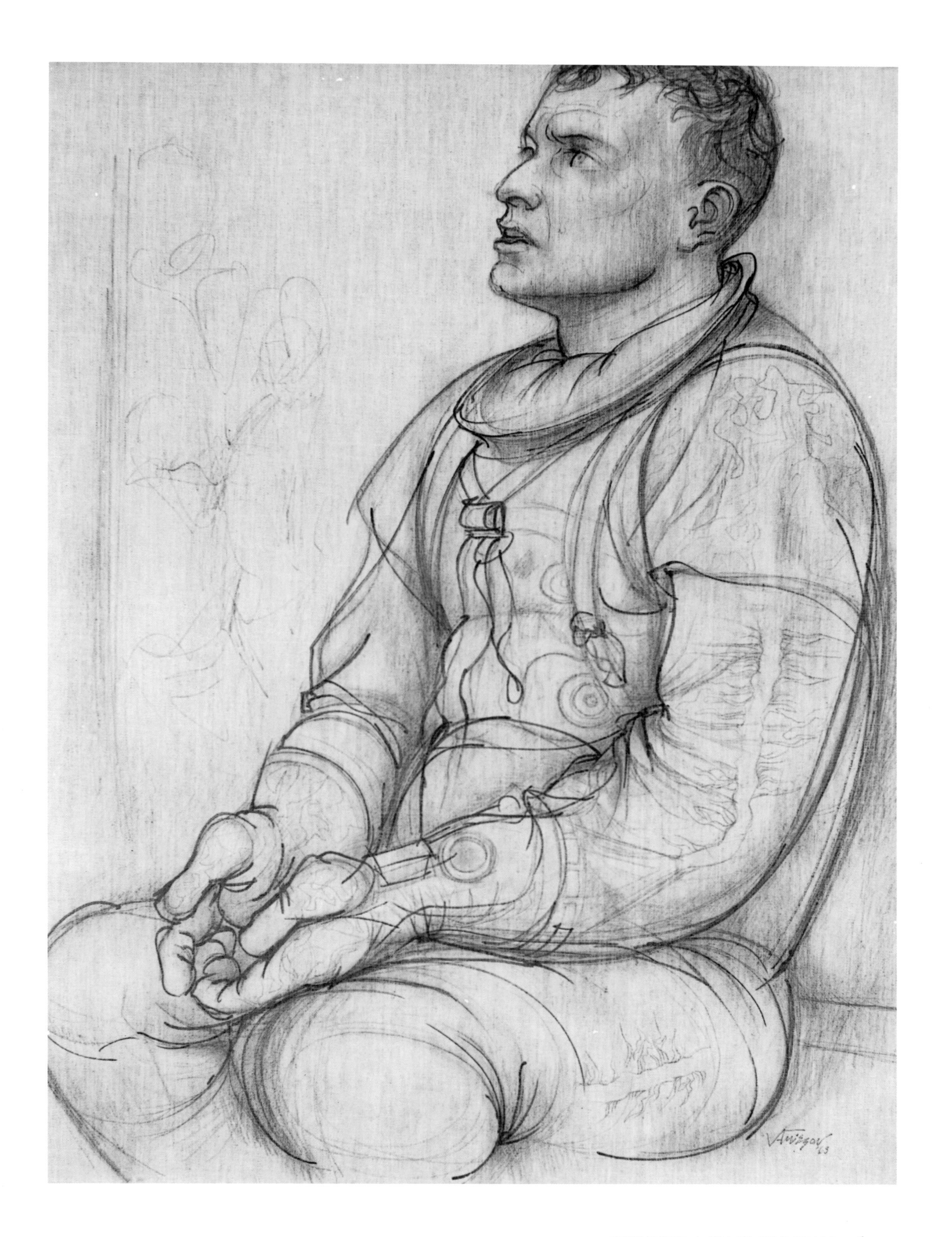

CHAPTER TWO

QUEST FOR THE MOON

UNITED
STATES

As president in 1960, John F. Kennedy regarded the U.S. space program less as a romantic vision of exploring the space frontier than as the practical application of a Cold War tool in maintaining the balance of power and spheres of influence in American-Soviet relations. Kennedy had been initially satisfied with allowing NASA to execute Project Mercury at a deliberate pace and to build on the satellite programs, which were yielding excellent results both in terms of scientific knowledge and practical application. In fact, during his March 1961 meetings with then–NASA administrator James E. Webb and Kennedy's budget director, David E. Bell, the president approved only a modest budget increase toward the development of big launch vehicles that would eventually support a moon landing.

Had the balance of power and prestige between the United States and the Soviet Union remained stable during the spring of 1961, it is quite likely that President Kennedy would have maintained this status quo and not advanced his moon program. American space exploration efforts might have then taken a radically different course. Two critical events, however, stepped up this nonchalant pace and changed the direction of U.S. space efforts.

On April 12, 1961 Soviet Cosmonaut Yuri Gagarin became the first human in space, making him a global hero and further demonstrating Soviet space capability. And although the United States managed to put Alan Shepard into space soon after, this effort was never quite enough to demonstrate U.S. technical equality with the Soviets. In the midst of a Cold War environment, this worried national leaders. Close in the wake of this achievement came another devastating event in the Cold War. Between April 15 and April 19, the Kennedy Administration had supported the aborted Bay of Pigs invasion into Cuba executed by Cuban refugees and aimed at overthrowing Fidel Castro. This debacle greatly embarrassed both the administration, and Kennedy personally, and damaged foreign relations.

A week after the Gagarin flight, Kennedy queried Vice President Lyndon B. Johnson—then head of the National Aeronautics and Space Council (NASC)—on putting together a strategy to catch up with the Soviets. A Kennedy memo to Johnson listed fundamental questions in that area, such as "Do we have a chance of beating the Soviets by putting a laboratory in space, or by a trip around the Moon, or by a rocket to go to the Moon and back with a man? . . . Is there any other space program that promises dramatic results in which we could win?" In response, Johnson recommended a U.S. commitment to a lunar landing—the best method of demonstrating the nation's technological prowess.

PAGES 66–67:
The Power to Go
PAUL CALLE
The five F-1 engines of the Saturn V rocket.

OPPOSITE:
Astronaut on the Moon
NORMAN ROCKWELL
Neil Armstrong and his historic first step on the Moon.

Cosmonaut Yuri Gagarin
PAUL CALLE

NASA's greatest consideration for fulfilling this committment was timing. Original estimates had projected a target date of 1967, but as the project began to crystallize, agency leaders recommended not adhering to such a strict deadline. Webb understood the problems associated with meeting target dates based upon NASA's experience with Mercury. He suggested that the president commit to a landing by the end of the decade, since that date would allow the agency another two years to solve any problems that might arise. The White House accepted his proposal.

President Kennedy's subsequent announcement of Project Apollo was made at a joint session of Congress on May 25, 1961, during which he stated: "If we are to win the battle that is going on around the world between freedom and tyranny, if we are to win the battle for men's minds, the dramatic achievements in space which occurred in recent weeks should have made clear to us all, as did the Sputnik in 1957, the impact of this adventure on the minds of men everywhere, who are attempting to make a determination of which road they should take . . ." He then added: "I believe this nation should commit itself to achieving the goal, before this decade is out, of landing a man on the Moon and returning him safely to Earth. No single space project in this period will be more impressive to mankind, or more important for the long-range exploration of space; and none will be so difficult or expensive to accomplish."

As demonstrated by this speech, Kennedy had accurately gauged the mood of the nation, utilizing it both politically and as an investment in the country's future. More important in his success, perhaps, was the unique and favorable confluence of conditions—political necessity, personal commitment and activism, scientific and technological ability, economic prosperity, and public mood—that created an exceptional window of opportunity, one which presents itself to national leaders only in the rarest of circumstances. The genius of Kennedy (and it is an open question that he fully grasped this fact) was to capitalize on an unusual consensus of mood among people, institutions, and interests which allowed the Apollo decision to be both made and accepted. It then fell to NASA to accomplish this most challenging of tasks—giving the American people a man on the Moon.

PROJECT GEMINI

Even as the Mercury program was underway and work was taking place to develop Apollo hardware, NASA program managers perceived a huge gap in the capability for human space flight that existed between that acquired with Mercury and what would be required for a lunar landing. They closed most of the gap by experimenting and training on the ground, but some issues called for experience in space. Three major areas arose. The first was the ability in space to locate, rendezvous, and join with another spacecraft. The second, closely related, was the ability of astronauts to work outside of a spacecraft. The third involved the collection of more sophisticated physiological data about human response to extended space flight.

What resulted was Project Gemini, a spacecraft built for NASA by McDonnell Aircraft Corporation, which flew with crews in 1965 and 1966. Accommodating two astronauts for flights of longer than two weeks in length, Gemini served well as the transition program between Mercury and Apollo.

When Thoughts Turn Inward
HENRY CASSELLI
Astronaut John Young is caught in a pensive moment prior to launch.

The capsule, launched by the Titan II rocket originally built as a ballistic missile for the Air Force, successfully met NASA's three requirements for space flight.

Although difficulties beset the program from the outset, the first operational mission took place on March 23, 1965, with Gemini 3. Mercury astronaut Gus Grissom commanded the mission, with John W. Young accompanying him. Spending five and a half hours in orbit, they utilized Gemini's thruster to manipulate their position and direction. The next mission, Gemini 4, was flown in June 1965 and lasted four days—the longest flight yet taken—and astronaut Edward H. White II performed the first extravehicular activity (EVA), or spacewalk, with the help of a handheld maneuvering apparatus. The Gemini 5 flight with Gordon Cooper and Pete Conrad flew in August and challenged the engineers at Houston's Manned Spacecraft Center with a duration of eight days. Gemini 7, carrying Frank Borman and Jim Lovell, subsequently pushed the envelope to two weeks, and on December 15, Gemini 6, with Wally Schirra and Tom Stafford, rendezvoused with Gemini 7, bringing yet another success and greater advancement for the program.

Despite the amazing achievements up to that point, there were no guarantees against failure, and one of the five Gemini flights that followed in 1966 reminded the agency that things could still go wrong. Gemini 8, with Neil Armstrong and Dave Scott as crew members, successfully linked up to an Agena target. Suspecting a failure in the target vehicle, Armstrong undocked, but one of Gemini's thrusters began firing on its own. The craft began spinning out of control, rotating once every second. Gemini 8 was threatening to break apart. In a quick move, Armstrong employed the thrusters designed for atmosphere reentry and stabilized the spacecraft. Although the mission ended prematurely, the astronauts splashed down safely in the Pacific Ocean.

Despite the disappointment, NASA worked hard to solve the thruster problem, in addition to other technological considerations. Mastering the

Five Point Motion Simulator
PAUL CALLE

BELOW LEFT AND RIGHT:
Gemini 6 Pilot Edward H. White II
PAUL CALLE
Astronaut Edward H. White during suit-up.

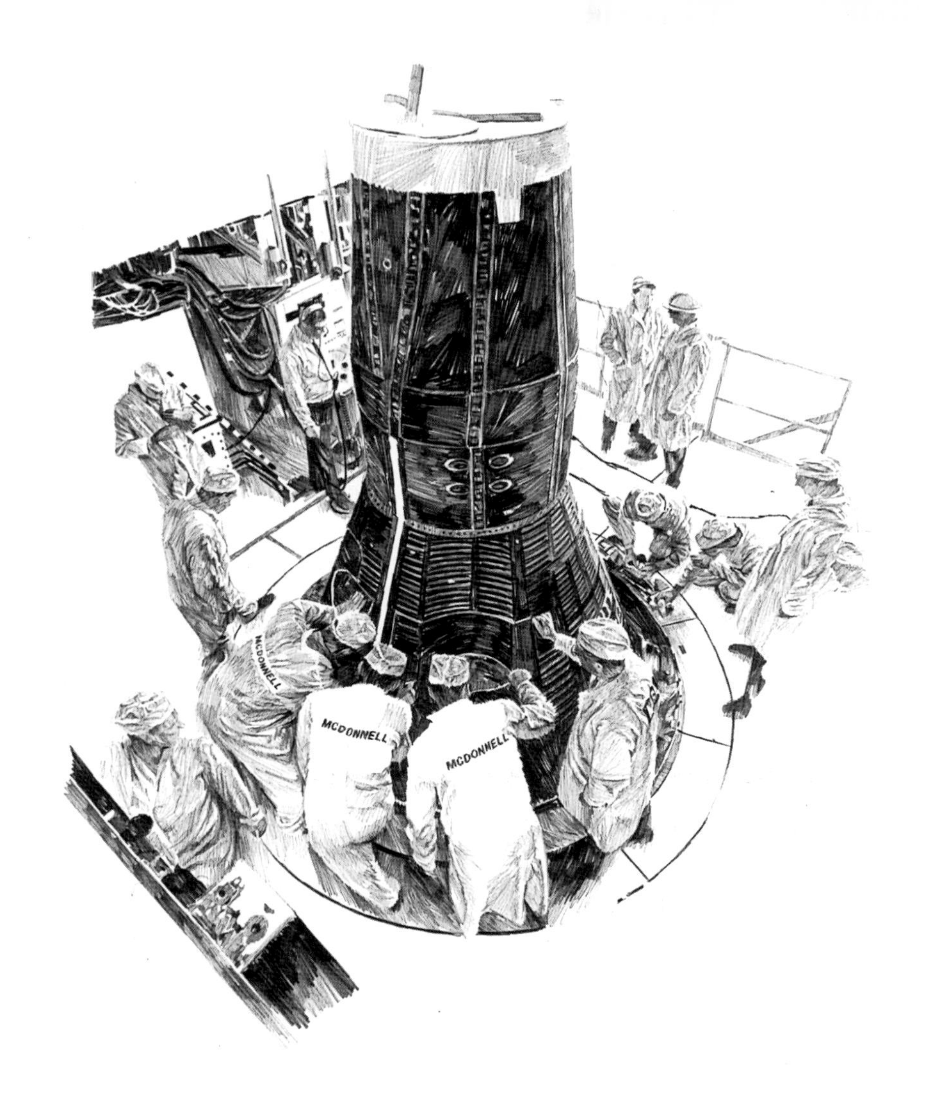

Gemini 6 in White Room During Countdown
PAUL CALLE

BELOW LEFT:
Gemini 4 Astronauts James A. McDivitt and Edward H. White at Complex 19
PAUL CALLE

BELOW:
Gemini 6 Astronauts Frank Borman and James A. Lovell, Jr. Walking Up Ramp at Complex 19
PAUL CALLE

ABOVE LEFT:
Recovery of Gemini 7
Astronauts Borman and Lovell
PAUL CALLE

ABOVE RIGHT:
Borman Boarding USS Wasp
PAUL CALLE

RIGHT:
Frank Borman and James A. Lovell, Jr.
PAUL CALLE

Rendezvous of Gemini 6 and 7
PAUL CALLE

The Bridge of Gemini 6 and 7 Prime Recovery Ship, USS Wasp
PAUL CALLE

Welcome Aboard
ROBERT MCCALL
Frank Borman and James A. Lovell, Jr. board the USS Wasp following their Gemini 7 splashdown.

OPPOSITE:
Moving Giant
JOHN W. MCCOY II
Launch-pad site of the ill-fated Apollo 1 mission.

spacewalk was a problem for Gene Cernan on the Gemini 9 mission; however, by Gemini 12, Buzz Aldrin had logged over five problem-free hours outside the spacecraft. With the power of the Agena rocket engine, Gemini 10 and 11 managed to increase altitudes to a record 850 miles.

Even before the final Gemini flight, NASA leaders had judged the program a huge success. Of the ten Gemini flights, six crews had rendezvous targets available in orbit and all six conducted successful rendezvous, a central goal of the program. The crews of Gemini 5 and 7 proved beyond any doubt that there were no physiological or operational barriers to the conduct of extended lunar missions. NASA also learned much about biomedicine during EVA, or extravehicular activity, and about the effects on the human body during extended periods of weightlessness. All this information was applied toward designing and executing the Apollo program, which was now ready for its operational phase.

THE APOLLO COMMAND AND SERVICE MODULE

Although much of the design work for Apollo had to take place prior to the knowledge gained in Gemini, the Apollo Command and Service Module

OPPOSITE:
Untitled
JAMES WYETH
Apollo astronauts working inside the Lunar Module Simulator.

(CSM) was based on features tested during the Mercury program and refined during the Gemini programs. In numerous ways the teardrop-shaped Apollo command module was a more complex and larger version of the conically shaped Mercury and Gemini spacecraft. Most important, it was a ballistic system, designed for orbital flight and return to Earth in exactly the same manner as its predecessors. Perhaps not surprisingly, the command module was the first Apollo flight hardware ready for flight testing.

The CSM, as eventually built, consisted of a three-person capsule capable of sustaining human life for longer than two weeks either in Earth orbit or in a lunar trajectory. It protected the astronauts from radiation and heat and served as the method of returning the astronauts to Earth following a completed mission. An attached cylindrical service module contained supplies, as well as the Service Propulsion System engine that placed the vehicle in and out of lunar orbit. The service module also provided stowage for most of the mission's consumables and remained attached to the command module during the flight to the Moon and the journey back to Earth, separating just before the Apollo capsule reentered Earth's atmosphere.

Work on the Apollo spacecraft began on November 28, 1961, with NASA awarding the prime contract for its development to North American Aviation. The capsule was ready for its first flight tests on May 13, 1964, when NASA launched a boilerplate model atop a stubby Little Joe II rocket. Early efforts proved successful—but just barely. Apollo spacecraft developers next tested the system, unmanned, in Earth orbit on September 18, 1964. By the end of 1966, after additional flight tests, the Apollo command module appeared ready for human occupancy.

Tragically, on January 27, 1967, the Apollo 1 launch-pad disaster took the lives of three astronauts. Gus Grissom, Ed White, and Roger Chaffee, who were aboard running a mock launch sequence, were asphyxiated in their sealed command module when a flash fire, aided by the pure oxygen atmosphere, broke out. Theirs were the first deaths directly attributed to the space program, and shock gripped the nation.

An eight-member investigation board, chaired by longtime NASA official and director of the Langley Research Center, Floyd L. Thompson, set out to discover the reasons for the tragedy. It was determined that a short circuit in the electrical system had caused the fire.

As a result, the first piloted flight test of the Apollo CSM was delayed for twenty months until mid-1968, when modifications in the CSM design, which eliminated the fire hazard and incorporated a number of essential upgrades, were completed. The unsung crew of Apollo 7 (Wally Schirra, Donn Eisele, and Walter Cunningham) then put the CSM through checks in low Earth orbit with an October 11-22 flight in 1968.

THE LUNAR MODULE

The Apollo 7 mission, as well as the Apollo 8 mission of December 1968, flew without the use of the lunar module (LM), then under development by the Grumman Aerospace Corporation. Begun a year late, in 1962, development of the lunar module always seemed to lag behind the aggressive schedule required to meet President Kennedy's mandate of reaching the Moon by the end of the decade. The Apollo lunar module was the first true spacecraft—

designed to fly only in a vacuum and devoid of the aerodynamic qualities that characterize Earth aircraft. Launched as an attachment to the Apollo Command and Service Module, it would separate from the CSM during lunar orbit and, with two astronauts aboard, descend to the Moon. Once on the lunar surface, it would serve as living quarters and base of operations. At the end of the astronauts' stay, the LM would fire its own rocket and rejoin the lunar-orbiting CSM. NASA's greatest concern for the lunar module was the need to devise two separate spacecraft components, one for descent to the Moon and one for the ascent back to the CSM.

THE SATURN ROCKET

Boosting the Apollo vehicles—the command and service module and the lunar module—beyond the Earth became the job of the giant three-stage Saturn V rocket, a member of the Saturn family of rockets, which were the brainchild of Wernher von Braun and the most powerful rockets yet developed.

The first of the Saturn rockets was the Saturn I, a research vehicle inherited by NASA when the Army Ballistic Missile Agency became a part of agency in 1960. Von Braun's engineers were hard at work on a launch vehicle that consisted of a cluster of eight Redstone boosters around a Jupiter fuel tank. Fueled by a combination of liquid oxygen (LOX) and RP-1 (a version of kerosene), the Saturn I's first stage could generate a thrust of 205,000 pounds. A second stage, known eventually as the Centaur, could generate another

OPPOSITE:
Test Stand
HOWARD KOSLOW
Saturn moon rocket engine tests at Huntsville, Alabama.

BELOW:
Engine Exhaust Bells
CHET JEZIERSKI
The thrusters of the Saturn V rocket.

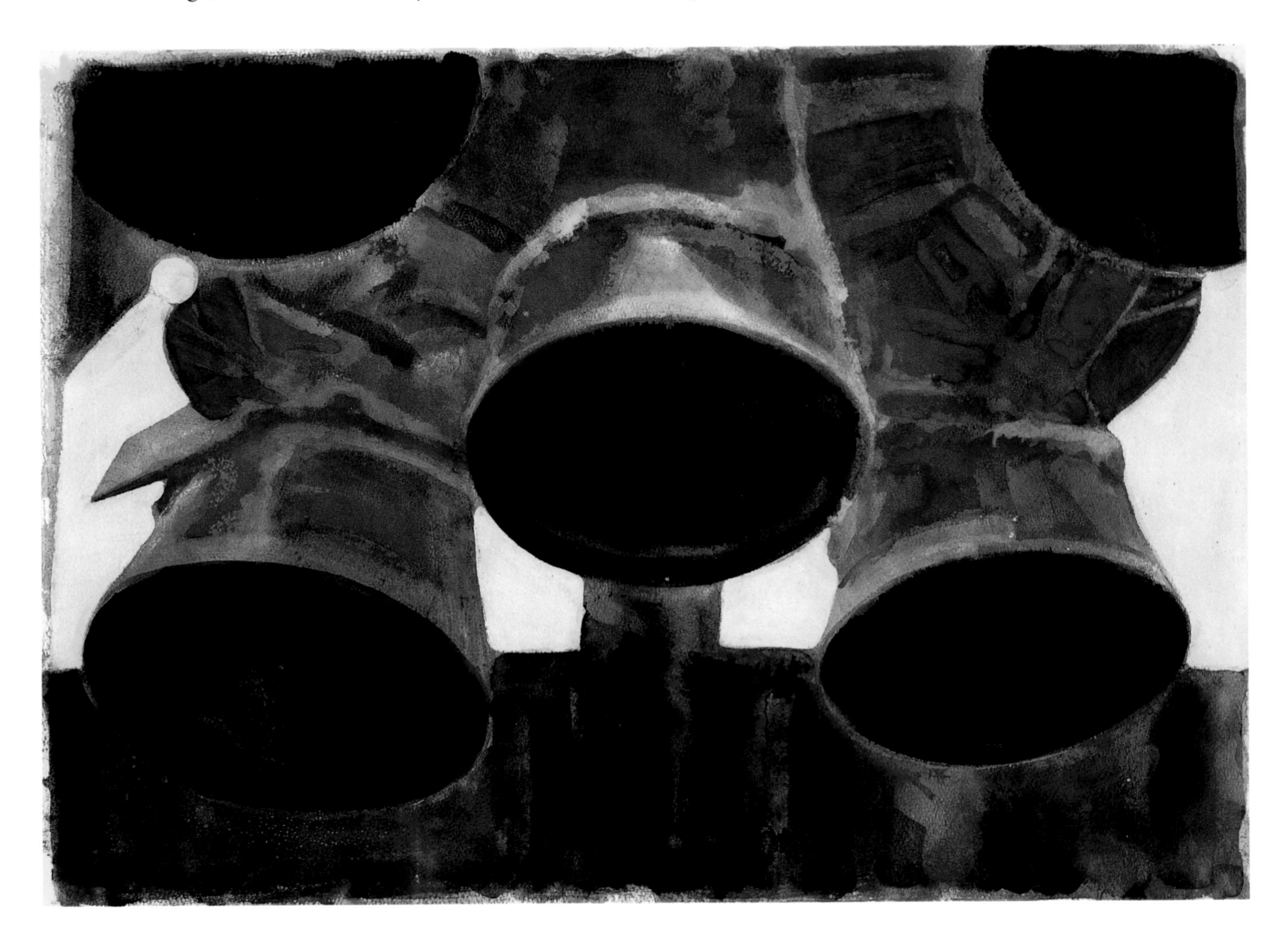

Flame Deflectors
PAUL SAMPLE

OPPOSITE:
Ladders
MITCHELL JAMIESON
Gantry, rocket, and flame detectors.

90,000 pounds of thrust. Flown only ten times between October 1961 and July 1965, the Saturn I demonstrated the technology necessary to move on to the next vehicle in the Saturn family of launchers.

The next step in Saturn rocket development came with the maturation of the Saturn I-B, an upgraded version of the earlier vehicle. With more powerful engines generating 1.6 million pounds of thrust from the first stage, the two-stage combination could place 62,000-pound payloads into Earth orbit. Flights with this launch vehicle tested the capability of the booster and the Apollo spacecraft in low Earth orbit (200–350 miles in altitude). On October 11, 1968, the Saturn I-B rocket was used to launch Apollo 7 and three astronauts into orbit as a final check to test the Apollo hardware that would be used for the flight to the Moon.

By far the largest launch vehicle in this family, the Saturn V represented the culmination of the earlier booster development and test programs. It was so large that it had to be assembled away from the launch pad and then transported to it. A fueled Saturn V weighed more than six million pounds at liftoff and, with the Apollo vehicles on top, stood 363 feet high. It featured three fuel-burning stages capable of powering astronauts to the Moon and, most important, bringing them back to Earth. The first stage generated seven and a half million pounds of thrust from five massive engines developed for the system. These engines, known as the F-1, were some of the most significant engineering accomplishments of the program, requiring the development of new alloys and different construction techniques to withstand the extreme heat and shock of firing. The second stage consisted of five engines that burned enough liquid oxygen and liquid hydrogen to deliver one million pounds of thrust. Because of liquid hydrogen's complexity as a propulsion fuel, the development of this stage always seemed behind schedule and required constant attention and additional funding to ensure completion.

A MAN ON THE MOON

Perhaps the most difficult decision of the entire Apollo program was the call made by George E. Mueller, head of NASA's Office of Manned Space Flight, to test the Saturn V in an "all-up" mode rather than test each of its components separately and incrementally. Predicated solely on the need to move the program forward quickly in the aftermath of the delays encountered during recovery from the tragic Apollo 1 fire, Mueller decided, in a calculated risk, to test the entire Apollo-Saturn system together in-flight without laborious preliminaries. This flew in the face of an incremental methodology advocated by von Braun that tested each component of every system individually and then assembled them for a long series of ground and flight tests.

Fire Damage
ALFRED MCADAMS

Apollo 8
Robert McCall

Mueller flew the overall Saturn V, along with the Apollo command and service module, on November 9, 1967, and again on April 4, 1968. These flights proved sufficiently successful for Mueller to declare the system ready for human occupancy. Mueller was prescient, for in seventeen tests and fifteen piloted launches, the Saturn booster family scored a one hundred percent launch-reliability rate.

With these successful test flights in low Earth orbit between 1967 and 1968, the Apollo program had reached a crossroads. Any continuation of Earth orbital missions would simply confirm what was already known. Equally important, the Soviets had sent the unmanned Zond spacecraft around the Moon—although not into lunar orbit—in September and November 1968 and had brought it back to Earth. The Zond missions were generally seen as precursors to a manned orbital mission. Accordingly, NASA officials agreed to make the Apollo 8 flight, scheduled for December 1968, a circumlunar one. Apollo 8 was therefore flown in part to prevent a Russian first that would have greatly diminished the psychological impact of any first landing by the United States. It might have been a landing mission except that the lunar module was still not ready; consequently, the closest that mission would come to the lunar surface was at an orbit of about fifty miles altitude. It was another bold decision made by the NASA leadership in an effort to meet the president's challenge.

On December 21, 1968, Apollo 8 took off atop a Saturn V booster from

THE VEHICLE ASSEMBLY BUILDING

Three times larger in volume than the Empire State Building and one of the largest buildings in the world, the cavernous Vehicle Assembly Building (VAB) at the Kennedy Space Center towers 525 feet above the flat Florida wetlands that surround it. Its completion in 1966, after four years of construction, marked a crucial milestone for NASA in the heat of America's race to the Moon. Today the VAB serves as a monument to America's accomplishments past, present, and future.

Columbia Rollout from the VAB, TRACY SUGARMAN

Activity in the VAB, HENRY PITZ

The building of this mammoth structure took tremendous effort. The drawings and specifications for it alone were so immense in their own right that a rail freight car was required to transport the necessary paperwork to interested bidders. Covering a ground area of about eight acres, its construction required more than 58,000 tons of steel, 45,000 separate steel pieces, and one million bolts. The foundation rests on 4,200 steel-pipe pilings driven to a depth of 160 feet. Equipped with two 325-ton cranes and 71 lifting devices, the VAB dwarfs huge rocket components, making them look like toy mobiles suspended from the corner of a playroom. A two-lane road, called the transfer aisle, runs through the building north to south and provides access to a myriad of motorized vehicles.

During Project Apollo, vertical stacking operations for the Saturn rocket stages required the efforts of hundreds of VAB personnel positioned in each of the four high bays. Against a backdrop of 456-foot high-bay doors, the incredible number of engineers and technicians working on the rockets would appear as if out of the pages of *Gulliver's Travels*. It was here in this hollow super-hangar that the most powerful staged rocket, the 363-foot-tall Apollo-Saturn V, was assembled and would go on to propel humankind into a new era of space exploration. Following the assembly and launch of the final Saturn I-B rocket in 1975, the VAB stood empty for five years until its next regular tenant—the space shuttle.

Nowadays, the VAB is the site where the shuttle orbiter, external fuel tank, and twin solid-rocket boosters come together in preparation for routine space-shuttle launches. Buildup operations begin with the transfer of solid-rocket booster segments from nearby facilities. After being hoisted onto a Mobile Launcher Platform in High Bays 1 or 3, they are mated to form complete boosters. Following inspection and checkout, the external tank is attached to the boosters. The orbiter, towed to the VAB from the Orbiter Processing Facility (OPF) by the Crawler-Transporter, is raised by overhead cranes and lowered onto the Mobile Launcher Platform, where it is mated to the boosters and external tank assembly. Once buildup operations are completed, the outer doors of the high bay open to allow the Crawler-Transporter to enter and move under the shuttle assembly, after which begins the journey to the launch pad, as well as the hope for yet another success to grace the pages of the VAB's already rich history.

JOEL WELLS, SPOKESPERSON,
KENNEDY SPACE CENTER

VAB, Nicholas Solovioff

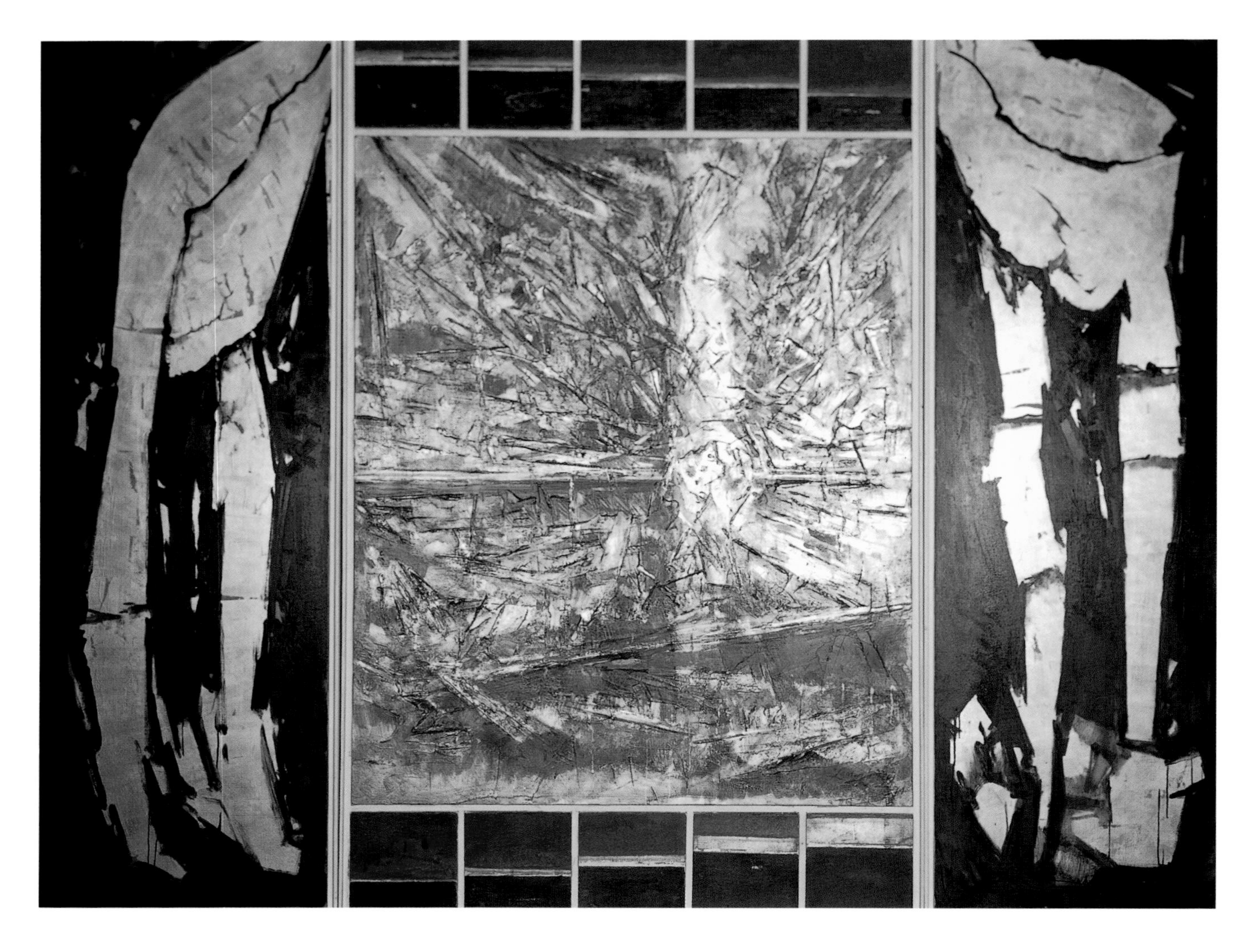

Apollo 9
Totality Triptych
LAMAR DODD

Kennedy Space Center with three astronauts aboard—Frank Borman, James A. Lovell, Jr., and William A. Anders—on its historic mission to orbit the Moon. The astronauts and ground controllers repeated all of the familiar phases: Earth orbit, circularizing the orbit, all as rehearsed. But then the Saturn third stage fired again and added the speed necessary for the spacecraft to escape Earth's gravity and place it on a trajectory to the Moon. On December 23, the three-man crew became the first humans to pass out of Earth's gravitational control and into that of another body in the solar system.

As Apollo 8 traveled outward, the crew focused a portable television camera on Earth and, for the first time, humanity saw its home from afar, a lovely and seemingly fragile blue marble hanging in the blackness of space. The crew also brought home photographic stills—stunning images of Earth rising above the horizon of the Moon. Apollo 8 also provided a remarkable opportunity to prove out flight-control procedures at lunar distances. It gave NASA vital experience in tracking a spacecraft orbiting the Moon and also afforded an opportunity to expand the photographic coverage of potential landing sites. It was a great confidence builder for the entire Apollo team, one shared with an international audience of hundreds of millions of people. Few who witnessed the event will ever forget the crew's Christmas Eve reading from Genesis that accompanied the televised pictures of the lunar surface passing below. It was a taste of more stunning moments and pictures yet to come.

Saturn I Booster, Marshall Space Flight Center
THEODORE HANCOCK

BELOW:
Inside the Vehicle Assembly Building at Kennedy Space Center
LOWELL NESBITT

Rollout
TRACY SUGARMAN

RIGHT:
Transporter
NICHOLAS SOLOVIOFF

OPPOSITE:
Parked Crawler
LESTER COOKE
Note the size of the individuals in and next to the Crawler

CRAWLER-TRANSPORTER

Golden slippers—that's how many refer to the metal tread links of the Crawler-Transporter. Of course, they aren't made of gold, and the "slipper" reference is a misnomer, since each plate is more than seven feet long and weighs 2,000 pounds. But back in 1965 the money spent on them inspired this nickname borrowed from an old song.

The Crawler-Transporter is a space-age colossus resembling a metal dinosaur. Between 1963 and 1966, the Marion Power Shovel Co. of Ohio was contracted to design and build two crawlers to transport the Apollo-Saturn V rocket to the launch pad at Kennedy Space Center. Today they transport the space shuttle.

Grasping the scale of a Crawler-Transporter is difficult because there is nothing comparable. Its design, based on Kentucky strip-mining tractors, features an enormous square platform about the size of a baseball infield, with each corner supported by eight gigantic double treads. Each tread boasts fifty-seven of the infamous "shoes." Checking in at six million pounds, the crawler is 131 feet long, 114 feet wide, 26 feet high, and gobbles a gallon of diesel fuel for every 35 feet it travels.

It's the crawler's job to move the two-story Mobile Launcher Platform, first to the Vehicle Assembly Building where the space shuttle is assembled atop the platform, and then out to the pad from where the shuttle is launched. Together, the crawler, Mobile Launcher Platform, and shuttle weigh eighteen million pounds. This moving skyscraper towers more than twenty-four stories high on a roadway specially designed to accommodate its massive load.

As huge as it may be, the precision with which the crawler operates is remarkable. Veteran operator Dick Beck, a mechanical engineer who has driven crawlers since the 1970's, says the crawler's error margin when unloading the MLP is a mere two inches. At the pad site, it must carry the platform and shuttle up a five-degree incline. Crawler motion, assembly height, roadway variations, and wind conditions are all factors which could affect the climb. Incredibly, the crawler's leveling system is so accurate that the tip of the shuttle's external tank doesn't vertically shift more than the diameter of a basketball.

"Just because the crawler inches along at a mere mile an hour doesn't mean it's an easy drive," cautions Beck. "It's like going to the store with a carload of kids—you have to be alert." Starting, steering, and stopping takes exertion and concentration. "It isn't like stopping your car," Beck says. "It takes about twenty to twenty-five feet to bring the crawler to a stop. All that kinetic energy has to dissipate. It's like stopping a semi-truck, only bigger."

Despite their slowness and the short distances they travel, the crawlers have logged over 1,400 miles. The Moon expeditions, Skylab, and every space shuttle flight to date have begun with a ride on the crawler. So perhaps there's a little gold in those "slippers" after all.

PAULA SHAWA, PUBLIC AFFAIRS WRITER, KENNEDY SPACE CENTER

Behind the Apollo 11
NORMAN ROCKWELL
Astronauts, scientists, engineers, astronauts' wives, and mechanics all look moonward toward destiny.

On that same Christmas Eve, Apollo 8 disappeared behind the Moon, where it would be out of radio communication with Earth. Not only were these three astronauts the first humans to see the mysterious back side of the Moon; while there, they had to fire the service module engine to reduce their speed enough to be captured into lunar orbit. At seventy miles above the surface, a breathtaking bird's-eye view of the battered lunar landscape was revealed. On Christmas Day they fired the service module engine once again, acquired the additional 3,280-feet-per-second speed needed to escape lunar gravity, and triumphantly headed back to Earth. They had confirmed, at close range, the lunar landing sites as being feasible and also proved hardware and communications performance at lunar distance. When the astronauts "splashed down" in the Pacific Ocean on December 27, 1968, a triumphant America warmly received the returning heroes.

Two additional missions, Apollo 9 and 10, occurred in the spring of 1969. Both were dress rehearsals for flights to the Moon, this time with the last all-important link, the lunar module, now available for use. Since the LM was like no other vehicle ever before flown, NASA officials and their astronauts wanted to test it thoroughly in the relatively safe confines of low Earth orbit. And because the LM could never fly in the Earth's atmosphere or gravi-

ty, few had enough confidence in ground or atmospheric tests of its capabilities to trust to it the lives of astronauts on the Moon. In Apollo 9 and 10, the LM went through its paces without incident. By the time of Project Apollo's end, the module had flown flawlessly in a series of nine missions, each more complex than the last.

The big event came with Apollo 11's liftoff on July 16, 1969. After confirmation that the hardware was working well, the three-day trip to the moon began. Once in lunar orbit, the crew checked out the Lunar Module Eagle, their precarious second home. On July 20, it separated and descended to the Moon's surface. At 4:18 P.M. EST, after more than six hours on the surface but still in preparation mode, the LM, with astronauts Neil A. Armstrong and Edwin E. "Buzz"Aldrin inside, landed on the lunar surface while Michael Collins continued to orbit overhead in Columbia, the Apollo command module.

"Houston, Tranquility Base here," the tinny radio crackled across more than a quarter of a million miles. "The Eagle has landed," said Neil Armstrong. It was a magnificent achievement, both technologically and managerially, and it demonstrated the dominance of the United States in space exploration. After checkout, Armstrong set foot on the surface, telling millions who saw and heard him on Earth that it was "one small step for man, one giant leap for mankind." (Armstrong later clarified this statement as "one small step for *a* man.")

Armstrong's crewmate, Buzz Aldrin, followed him out onto the lunar surface and captured the essence of the experience in the phrases: "Beautiful, beautiful. Magnificent desolation." He later observed "the contrast between the starkness of the shadows and the desertlike barrenness of the rest of the surface. It ranged from dusty gray to light tan and was unchanging except for one startling sign: our LM sitting there with its black, silver, and bright yellow-orange thermal coating shining brightly in the otherwise colorless landscape."

Armstrong and Aldrin planted an American flag (without claiming the land for the United States), unveiled a plaque with the inscription "We come in peace for all mankind," collected soil and rock samples, and set up

Suiting Up
PAUL CALLE
Neil Armstrong, Edwin Aldrin, and Michael Collins.

RIGHT:
Apollo 11 in the VAB
MARIO COOPER

OPPOSITE:
Stacking the Saturn
FRANKLIN MCMAHON

tific experiments. A few hours later the astronauts returned to the Eagle and blasted back to the Command Service Module Columbia in orbit overhead. The next day the crew launched back to the Apollo capsule and began the return trip to Earth, splashing down into the Pacific Ocean on July 24.

With Apollo 11, an eight-year national commitment had been fulfilled; Americans were on the Moon. More than one-fifth of the Earth's population had seen the ghostly television images of the two space-suited men as they gingerly went about their work in the unlikely world of gray surface, boulders, and rounded hills. The astronauts returned to an ecstatic reception reminiscent of the excitement felt in the early 1960's for John Glenn and the Mercury astronauts. For a brief moment, people's day-to-day divisions were suspended; the world watched and took joint pride in this singular achievement.

During a three-year period between 1969 and 1972, the United States completed five additional moon-landing missions. The final Apollo mission—Apollo 17, launched in December 1972—perhaps best encapsulated the accomplishments and meaning of Apollo. Its astronauts brought back from the moon a treasure of 242 pounds (110 kilograms) of lunar material, 2,100 photographs, and twenty-two hours of experience scouring the Taurus-Littrow Basin for answers to questions almost as old as the human race.

Interpretation: Apollo 11 Moon Landing
Tom O'Hara

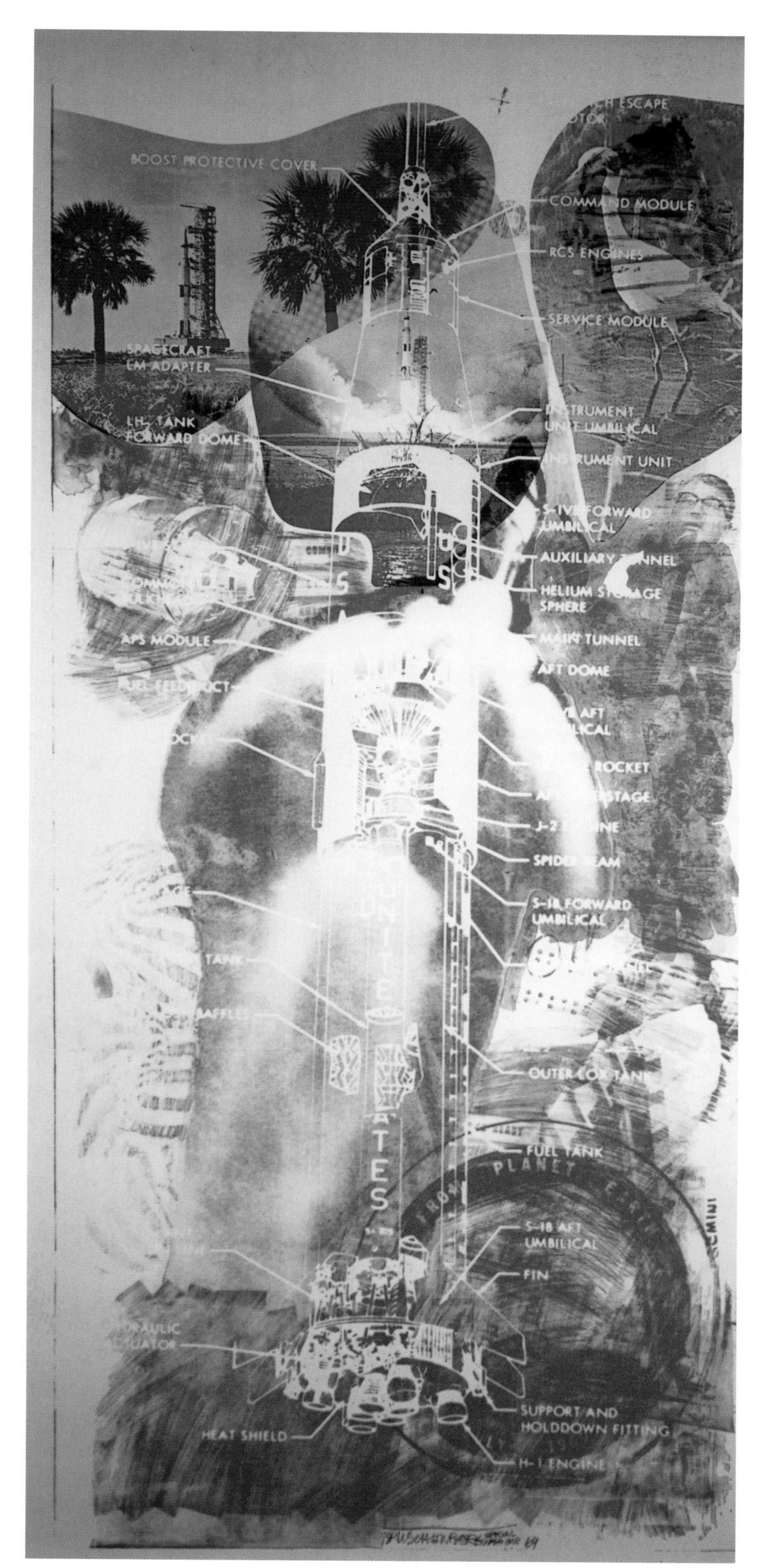

Sky Garden
ROBERT RAUSCHENBERG

CAPCOM Bruce McCandless Communicates with Apollo 11
FRANKLIN MCMAHON

Gene Cernan, Apollo 17's commander and the last human to leave a footprint on the lunar surface, summarized his feelings on Apollo by remarking how "we examined the rewards, the worth of our risks, managed our risks, got as smart as we possibly could, and worked hard to get smarter, and committed ourselves to doing something we believed in and everything it took to support that commitment."

Harrison Schmitt, Cernan's fellow crew member and the first geologist to do field work on the Moon, considered this final mission as the capstone of a great set of voyages of discovery. During the astronauts' return to the command module from the lunar surface, President Richard Nixon had issued a

Apollo 11 Module Being Hauled Aboard the USS Hornet, November 24, 1969
PAUL D. DRILLIP

Recovery Remembered—Apollo 16
CHET JEZIERSKI

OPPOSITE:
Rendezvous, Apollo 11
ROBERT MCCALL
The Eagle rejoins the Command Service Module Columbia in orbit.

ABOVE LEFT:
Head Phones
JOSEPH C. CHIZANSKOS

ABOVE RIGHT:
Mission Control
JOSEPH C. CHIZANSKOS

LEFT:
Communicating with Astronauts
JOSEPH C. CHIZANSKOS

OPPOSITE LEFT:
Hey Houston, We've Got a Problem!
JOSEPH C. CHIZANSKOS
These startling words set off an uncertain journey home.

OPPOSITE RIGHT:
Three Astronauts Inside the Apollo 13
JOSEPH C. CHIZANSKOS
Astronauts Lovell, Haise, and Swigert huddle to preserve body heat.

APOLLO 13

Apollo 11, the first lunar landing flight, honored President Kennedy's commitment to land men on the Moon and bring them back safely by the end of the decade. Since this flight was intended as a technical challenge and not a scientific expedition, a flat area or "sea" of the Moon was selected for the landing to ensure the maximum chance for success. Apollo 12 repeated this challenge, and it, too, landed in a lunar sea.

Following these two flights, however, the scientific community wanted to be heard. Scientists had acquired enough rock samples from the Moon's seas and were now interested in rocks from its hills or highlands. The reflectivity from these areas was brighter; therefore, the material could be different, and it was known that ejecta from volcanism or impacts littered the surface. Such samples could provide clues to the Moon's interior. Consequently, Apollo 13 was assigned a different objective for its Moon mission—to land in the hills surrounding a crater named Fra Mauro. Crew training, involving astronauts Lovell, Haise, and Swigert, was to concentrate on lunar surface exploration. The adopted motto, "Ex Luna Scientia" (knowledge from the Moon), clearly defined this mission.

All of the careful preparations made toward gaining greater knowledge of the Moon's origin came to naught, however, when on April 13, 1970, the moonbound Apollo 13 experienced an oxygen-tank explosion of such great magnitude as to cripple the Command Module Odyssey seriously and place the crew in mortal danger.

The four days following the explosion remain a monumental testament to crisis management. The crew at Mission Control in Houston, working closely with the astronauts, determined solutions to each crisis as it arose. In order to return the spacecraft to Earth before its batteries expired, a new course had to be determined and a "speed up" plan worked up. The Commander had to learn, once again, how to control the Lunar Module since it had never been maneuvered with the Command Service Module still attached. Inspiration came to the forefront when crew-system personnel devised a solution to remove the excess carbon dioxide that was threatening the crew's survival.

Slowly, the odds of a safe return increased as one crisis after another was solved. The leadership, teamwork, initiative, and motivation of a group of "can do" individuals turned an almost certain catastrophe into a safe recovery. Failure was never an option with the Apollo 13 team.

In retrospect, if there was to have been an explosion on the way to the Moon, it could not have happened at a better time. If it had taken place early in the flight, the batteries would have died before the spacecraft could reach the Earth; if the explosion had occurred after the spacecraft entered lunar orbit, there would not have been sufficient Lunar Module fuel to return the crew home.

Yes, the flight of Apollo 13 was a failure—but a successful failure. It will go down in history not as a significant addition to the knowledge about the Moon, but to the knowledge of what human beings are capable of doing to achieve victory in the face of certain danger.

CAPTAIN JAMES A. LOVELL, JR.

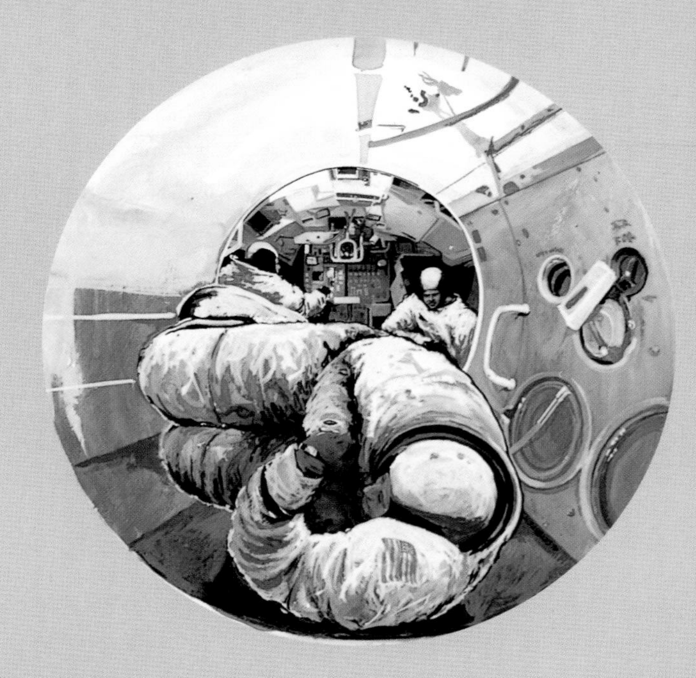

Station 13
APOLLO 16
FROM LIVE TV
april 23 1972

NAVY
67
USS PRINCETON

Photographers Shoot Apollo 17
NICHOLAS SOLOVIOFF

OPPOSITE TOP:
Live EVA Broadcast, Apollo 16, April 23, 1972
TOM O'HARA

OPPOSITE BOTTOM:
Aboard the USS Princeton
TOM O'HARA
Helicopters are being prepared for recovery activities.

statement praising the accomplishments of Project Apollo but also, in effect, stating that the Apollo 17 crew would be the last to visit the Moon that century. "I remember that upsetting me greatly," said Schmitt. "One, because I thought it was an unnecessary remark even if it was true, and two, I hoped that we would prove him wrong," Mission Controller Eugene Kranz agreed. "I would say it's the damnedest disappointment. . . . It was almost in a state of sorrow that we lifted off for that last time with the crew from the surface. When we brought that LM up, it was almost a time of mourning."

Project Apollo in general, and the flight of Apollo 11 in particular, should be viewed as a watershed in the nation's history. It was a feat that demonstrated both the technological and economic virtuosity of the United States and established scientific preeminence over rival nations—the primary goal of the program when first envisioned by the Kennedy administration in 1961. It had been an enormous undertaking, costing $25.4 billion (about $99 billion in 1998 dollars), with only the building of the Panama Canal rivaling the Apollo program's size as the largest nonmilitary technological endeavor ever undertaken by the United States; only the Manhattan Project compared in a wartime setting.

Several important conclusions have come out of Project Apollo. First, the Apollo program succeeded in accomplishing the political goals set for it as a direct result of the 1961 Cold War crisis brought on by several separate factors, not the least of which was the Soviet orbiting of Yuri Gagarin and the disastrous Bay of Pigs invasion.

Second, Project Apollo was a triumph of management in fulfilling

enormously difficult engineering, technological, and organizational system integration requirements. James E. Webb, the NASA administrator at the height of the program between 1961 and 1968, always contended that more than anything else, Apollo was a management exercise and that the technological challenge, while sophisticated and impressive, was largely within grasp at the time of the 1961 decision to begin the program. NASA leaders had to acquire and organize unprecedented resources to accomplish the task at hand from both a political and technological perspective. Webb's continual maneuvers in Washington ensured the sufficient resources necessary to meet Apollo requirements; NASA personnel managed the complex structures of a multifarious task by employing a "program management" concept that centralized authority and emphasized systems engineering.

Third, Project Apollo forced the people of the world to view the planet Earth in a new way. Apollo 8 was critical to this fundamental change since it had treated the world to the first pictures of the Earth from afar. The poet Archibald MacLeish summed up the feelings of many people when he wrote at the time of Apollo that "to see the Earth as it truly is, small and blue and beautiful in that eternal silence where it floats, is to see ourselves as riders on the Earth together, brothers on that bright loveliness in the eternal cold—brothers who know now that they are truly brothers." The modern environmental movement was galvanized in part by this new perception of the planet and the need to protect it and the life that it supports.

Finally, the Apollo program, while an enormous achievement, left a divided legacy for NASA and the aerospace community. The perceived golden age of Apollo fostered in the agency the expectation that the direction of any major space goal would always find a broad consensus of support and provide NASA with adequate resources and the license to dispense them as it saw fit. Something most NASA officials did not understand at the time, however, was that Apollo had been conducted under exceptional political circumstances that would not be repeated.

The Apollo decision was, therefore, an anomaly in the national decision-making process and a dilemma that was difficult to overcome. However, in the recent past, moving beyond the program to embrace future opportunities has been an important goal of the agency's leadership. Exploration of the solar system and the universe remains as enticing a goal and as important an objective for humanity as it has ever been. Project Apollo was an important early step in that ongoing process of human exploration.

PAGES 106-107:
The Last Apollo
NICHOLAS SOLOVIOFF
Apollo 17, the final manned expedition to the Moon.

The last Apollo Dec 72

CHAPTER THREE

FROM COMPETITION TO COOPERATION

Since the beginning of the space agency itself, foreign relations have been an inextricable element in the fabric of the U.S. space program. From the Mercury and Apollo programs to NASA's space-shuttle efforts and its plans for future space stations, the furtherance of foreign relations has been and continues to be a powerful shaper of policy-making and project execution in U.S. human space-flight endeavors.

During its first decade and a half, however, the space program was dominated by a foreign-relations policy that focused on international rivalry and the quest for world prestige. The quintessential example of the power struggle between the Soviets and Americans—the race to the Moon—was an intensely competitive challenge between two Cold War superpowers seeking to outdo each other in a symbolic battle of ideologies. The balance was forever shifted in 1969 when the astronauts planted the American flag on the Moon's surface. The irony of planting that flag, coupled with the statement that "We come in peace for all mankind," was not lost on Soviet leaders. They realized that in that particular context, they were not considered a part of "all mankind."

Following this coup and the successful completion of the Apollo program, it seemed pointless to continue in an atmosphere of international competition; the United States was now the undisputed world leader in scientific and technological virtuosity. Yet, although President Kennedy's objectives were duly accomplished with technical brilliance, the legacy of the Apollo program held no logical next step for many American people—they were indifferent. The world for them had changed, and the country's attention had turned to the tumultuous political, economic, and social eruptions of a new era. The advent of inflation, civil rights activism, turmoil over the Vietnam War—these represented the country's immediate concerns. And as president during this time, Richard M. Nixon made it clear that there would be no further Apollo-like space efforts during his leadership.

The NASA budget and personnel ranks began to decline, and in order to survive the cuts, the agency and those working for a continued aggressive U.S. space-exploration effort developed a new model: cooperation with allies rather than competition with the nation's Cold War rival. It seemed the only way to maintain a link between space exploration and foreign-relations objectives. From the 1970s on, all subsequent major human space-flight efforts, as well as later minor projects, became identified with international partnerships.

PAGES 108–09:
Morning Launch of the Challenger (STS-7), June 18, 1983
GREG MORT

OPPOSITE:
Space Telescope in Orbit
PAUL HUDSON

THE APOLLO-SOYUZ TEST PROJECT

Apollo-Soyuz—the final flight of the Apollo program—served as the first international human space flight and took place at the height of the U.S.-Soviet détente. Although begun as a diplomatic effort toward a thaw in relations between the two superpowers, the Apollo-Soyuz Test Project (ASTP) had ambitious technological goals as well. It was designed to test the compatibility of rendezvous and docking systems for American and Soviet spacecraft.

For the project, the United States incorporated an Apollo command and service module (CSM) modified to support mission experimentation. To enable a docking between the dissimilar spacecraft, extra propellant tanks, controls, and equipment related to the docking module were jointly designed by the United States and the Soviet Union, then built in the States. In a further sharing of experience and technology, the astronauts and cosmonauts repeatedly visited each others' space centers; in April 1975,

Apollo-Soyuz Spacecraft
PAUL CALLE

OPPOSITE:
Docking Module Mock-up
TOM O'HARA

PAGES 114–15:
Apollo-Soyuz Test Project
ROBERT MCCALL
The docking of Soviet and American spacecraft marked a thaw in Cold War diplomacy and heralded a new era of international cooperation.

Tom O'Hara

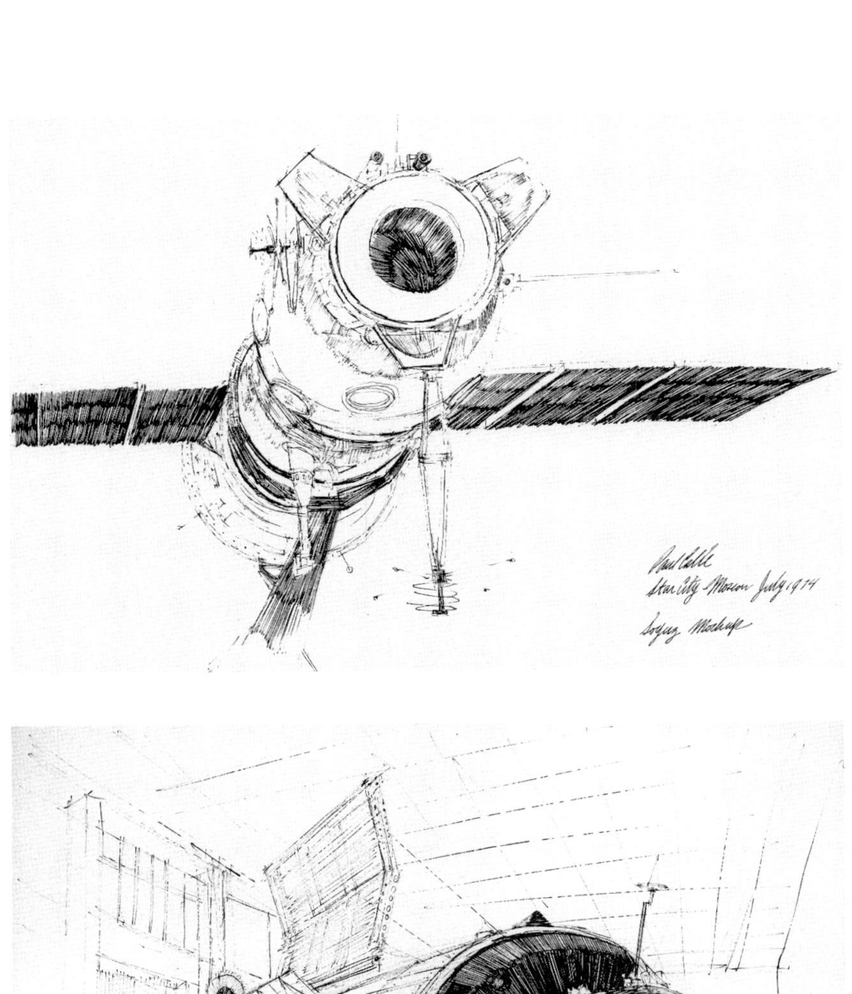
Paul Calle
Star City - Moscow July 1974
Soyuz Mockup

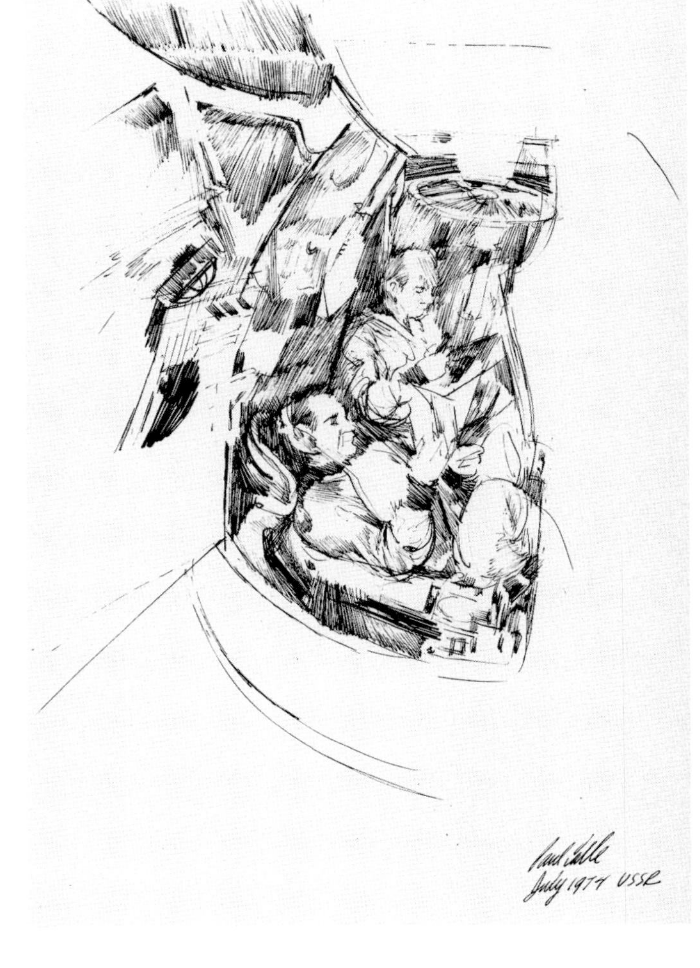
Paul Calle
July 1974 USSR

Paul Calle Moscow July 1974

CCCP
Paul Calle
Moscow June 1974

Paul Calle July 1974 USSR

U.S. astronauts became the first Americans to see the Russian launch facilities at Tyuratam.

On July 15, 1975, at 5:20 P.M. Russian local time, Soyuz 19, carrying cosmonauts Aleksey A. Leonov and Valery N. Kubasov, was launched from Baykonur Cosmodrome. The Apollo launch took place seven hours and thirty minutes later from Kennedy Space Center with astronauts Thomas P. Stafford, Vance D. Brand, and Donald K. "Deke" Slayton aboard. On July 17 at 8:51 A.M. EDT and after a series of Apollo maneuvers, Brand reported sighting Soyuz and made voice contact soon after. Successful docking of the two spacecraft came at 12:12 P.M. on July 17 over the Atlantic Ocean, and three hours later the two crews opened their hatches and Stafford and Leonov shook hands. "Glad to see you," Stafford told Leonov in Russian. "Glad to see you. Very, very happy to see you," Leonov responded in English.

Both Soviet Communist Party General Secretary Leonid I. Brezhnev and President Gerald R. Ford congratulated the crews and expressed their confidence in the success of the mission. Stafford then presented Leonov with "five flags for your government and the people of the Soviet Union" with the wish that "our joint work in space serves for the benefit of all countries and peoples on the Earth." Leonov presented the U.S. crew with Soviet flags and plaques. The men signed international certificates and exchanged other commemorative items.

For the next three days, in addition to carrying out their own mission experiments, the crews conducted joint activities, including visiting each others' spacecraft, sharing some meals, videotaping scientific demonstrations, and giving American and Russian television viewers tours of the spacecraft. Additional speeches and exchanges of commemorative items were made before both U.S. and Soviet audiences before the final handshakes took place on July 18 at 4:49 P.M. EDT, at which point the crews returned to their respective spacecraft. Brand wished Leonov and Kubasov success, and the hatches were closed. The Apollo and Soyuz spacecraft undocked on July 19, 1975, and

OPPOSITE PAGE:
Sketches by Paul Calle made in July of 1974 in Star City, Moscow.

Star City, Moscow
PAUL CALLE
Astronaut Deke Slayton and cosmonaut Alexei Leonov confer on mission procedures.

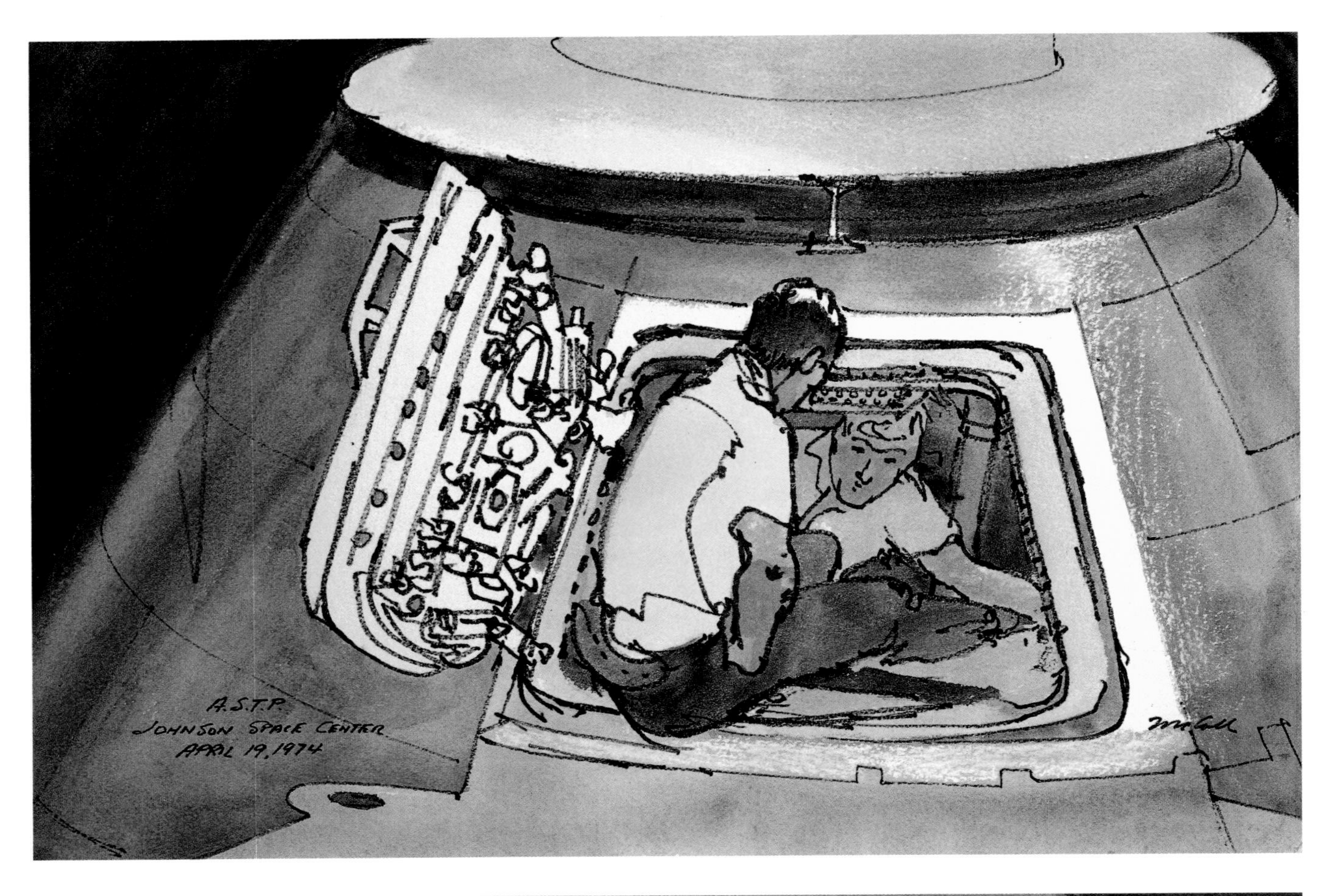
A.S.T.P.
JOHNSON SPACE CENTER
APRIL 19, 1974
McCall

A.S.T.P.
McCall
4/19/74
J.S.C.

as the spacecraft departed, Leonov thanked Apollo for a job well done and "a very good show." Brand returned the sentiment.

Soyuz 19 remained in orbit until July 21, when it returned safely to Earth after a 142-hour, 31-minute mission. Apollo remained in orbit while its crew completed science experiments, then splashed down into the Pacific Ocean on July 24, marking the last-ever ocean landing for U.S. space flights.

As a cooperative mission, ASTP proved enormously successful. The handshaking aboard the docked spacecraft symbolized détente between the Cold War rivals—a significant goal for the Nixon and Ford administrations. As a scientific mission, ASTP was less notable, but it did provide some scientific return in astronomy. Most of all, however, the mission demonstrated the close ties between foreign relations and space exploration. The lesson of cooperative venture was not lost on either American or other national leaders, sounding instead a persistent theme in subsequent space exploration.

OPPOSITE TOP:
Apollo-Soyuz Test Project, Johnson Space Center
ROBERT MCCALL

OPPOSITE BOTTOM:
Docking Module (Interior) Apollo-Soyuz Test Project
ROBERT MCCALL

PAGES 120–21:
The Spirit of Discovery
TERRY WOFFORD
The Space Shuttle Discovery sits on the launch pad in a prismatic kaleidoscope of light.

THE SPACE SHUTTLE

Few people view the reusable space shuttle—presently the most visible ongoing program of NASA—as an international project. Even so, it owes its existence to a policy of international cooperation and has enjoyed a measure of global participation right up to the present.

Begun as a project in February 1969, while Project Apollo was underway and Richard M. Nixon had entered office, the space shuttle was originally designed as an integral part of a large program to establish a space station. Its goal was to provide cost-effective "routine access to space." With the hopes of an expansive post-Apollo program dashed by Nixon's budgeteers, however, NASA's administrator Thomas O. Paine had to find a way to maneuver around the resulting fiscal barrier and secure necessary funding.

Nixon, whose administration enjoyed enormous success in foreign relations, naturally gravitated to projects which furthered the position of the United States among the nations of the world. Paine played on the administration's desire to extend international involvement in U.S. activities by offering to make the shuttle an international program. Paine explained to Nixon in February 1969 that European space activities had been limited thus far to communication satellites, but that opportunities for greater partnership were present in post-Apollo activities. He believed that the United States and the European nations could help solidify NATO alliance and strengthen it through cooperation in space projects.

After receiving a positive response, Paine sought European support for the shuttle. In doing so, he knew that the more partners NASA had in the venture, the greater was the likelihood of obtaining approval for the project. On August 11, 1969, Paine presented the shuttle program to Professor Hermann Bondi, Director-General of the European Space Research Organization (ESRO). Paine proposed that NASA and European space-faring nations previously excluded from Project Apollo now join forces in a space-shuttle program that held NASA's faith as the technology of the future. To entice the Europeans, Paine's proposal included the flying of European astronauts and the possibility of participation in the construction of this new and farsighted launch system. Europeans could build modules that would fly on it and, in the process, leverage their own technological capability.

USA
NASA
Discovery

NASA
United States

Thomas Paine then tried to enlist the support of Canada, Australia, and Japan. He visited Canada in December 1969, and took a trip to the Pacific between February and March 1970. He outlined NASA's post-Apollo plans to senior space officials and asked for cooperation. Paine argued that President Nixon was committed to reducing Cold War tensions between the United States and the Soviet Union and that international rivalry was not the focus of his space program—rather, the president was anxious for an international partnership.

The response was positive but noncommittal. Within months, however, the Australians and the Japanese had withdrawn from the project and decided to place emphasis on space activities that served their own national interests, although they maintained interest in purchasing shuttle services from NASA. The Canadians decided to participate in the shuttle program and ultimately developed the Canada arm remote manipulator crane currently being flown on the shuttle.

During this time NASA was careful over how partnerships would be negotiated. The U.S. agency had several goals in handling this issue. It wanted European involvement, without question. It was not lost on Paine and other key NASA leaders that every partnership gained brought greater legitimacy to the overall program and thus brought the program closer to presidential approval, but negotiations had to be done on NASA's terms. Although probably overlooked by NASA at the time, agreements with foreign nations could also protect the shuttle program from drastic budgetary and political changes. U.S. politicians and diplomats would not relish any serious international incident resulting from changes to the shuttle's technological program, such as altered funding, scheduling, or other factors in response to short-term budgetary needs.

Although NASA leaders were unusually unified in support of the positive political results that could come with European partnership, they recognized several drawbacks as well. Assigning an essentially equal partner the responsibility for developing a critical subsystem meant giving up the power to make changes, dictate solutions, and control schedules and other factors. In addition to this concern, some technologists expressed fear that bringing Europeans into the project really meant giving foreign nations technical knowledge heretofore held only by the United States. No other nation had built a reusable spacecraft before, and only three had launch capability. Some NASA officials, but even more State Department personnel, questioned the prudence of ending such a technological monopoly.

Concerns of technology transfer remained unsettled until the fall of 1971, when the stage was set for the development of the space shuttle. NASA officials had been working hard within the Nixon Administration to achieve formal program approval, but the Office of Management and Budget (OMB) was instead considering further NASA budget cuts, including halting space-shuttle development and canceling Apollo 16 and 17. In August 1971, Caspar W. Weinberger, Deputy Director of the OMB, wrote a memo to the president, arguing that "there is real merit to the future of NASA, and to its proposed programs." He felt cuts would cause those at home and abroad to believe that "our best years are behind us, that we are turning inward, reducing our defense commitments, and voluntarily starting to give up our superpower status, and our desire to maintain world superiority." This perspective—along

OPPOSITE:
Columbia at Booster Separation
ROBERT MCCALL
Solid rocket boosters separate from the orbiter and external tank during the first flight of the Space Shuttle Columbia, April 12, 1981.

OPPOSITE:
Columbia Launch Fantasy
ANDREAS NOTTEBOHM

with the fact that a new NASA project would aid unemployment in states important to the 1972 presidential election—ultimately held more weight than whatever effect European involvement would have had on the program. In a handwritten scrawl on Weinberger's memo, President Nixon replied, "I agree with Cap," and, in January 1972, gave the go-ahead for the space-shuttle project.

The space shuttle that emerged from this process was a partially reusable vehicle that would be launched atop an expendable booster, while a second stage—an orbiter about the size of a Boeing 707—flew on under its own power into orbit, performed its mission, and then returned to Earth. The orbiter would be reusable, the boosters partially so. Everyone believed, and the White House's press announcement stated, that the shuttle could significantly reduce both development and operation costs, making the system attractive both to administration officials concerned with the long-term investment and users who wanted to reduce the cost of space flight.

Within two weeks of this announcement, and still overcome with euphoria at the space-shuttle decision, Nixon's newly appointed NASA administrator, James C. Fletcher, readdressed the issue of European collaboration in building the system. The Europeans were elated over the program approval; it had cleared away the question of long-term U.S. commitment. In November 1972, a major collaborative effort between the U.S. and European space leaders led to the development of a science module that would fit into the shuttle cargo bay—a sortie laboratory, later known as Spacelab. It would function as a major payload facility for onboard shuttle use and permit foreign astronauts to participate in flight and conduct their own experiments. Its development by the European Space Agency represented a $250 million contribution to the overall space effort.

There was tremendous excitement when Columbia—the first orbiter to fly in space, and the first space vehicle that did not undergo unmanned test flights—took off from Cape Canaveral, Florida, on April 12, 1981, piloted by astronauts John W. Young, a spaceflight veteran, and Robert L. Crippen, a key member of the shuttle-flight test team that conducted atmospheric tests on the very first space orbiter, the Enterprise.

It was the Columbia launch that renewed high hopes for the achievement of low-cost, routine access to space. Thereafter, one successful, spectacular mission after another took place and NASA and the European partners made good use of the new capability. Though the most important mission of any spacecraft designed to carry astronauts was to ensure crew safety, the space shuttle was used as a platform for a number of important scientific endeavors.

Yet, despite these high hopes, the shuttle program ultimately provided neither inexpensive nor routine access to space. By January 1986, there had been only twenty-four shuttle flights, and the vehicle itself was neither cheap nor reliable. Although the system was reusable, its complexity, coupled with the ever-present rigors of flying in an aerospace environment, meant that the turnaround time between flights was several months instead of several days.

It was the tragedy of the Space Shuttle Challenger (STS-51L), launched from Kennedy Space Center on January 28, 1986, that subsequently brought the shuttle program under even harsher scrutiny. Seventy-three seconds into the flight, an explosion occurred as a result of a leak in one of two solid-

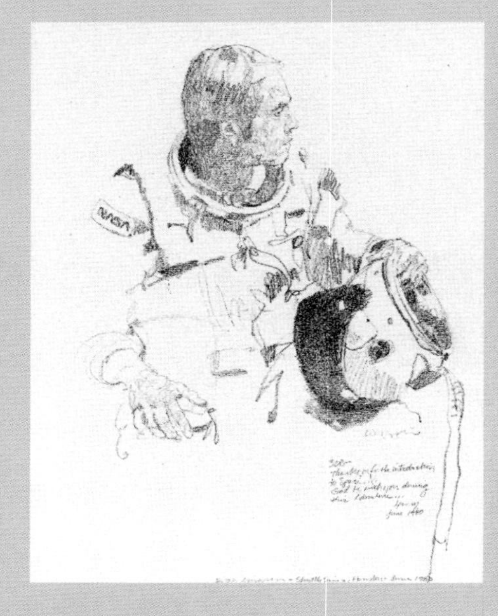

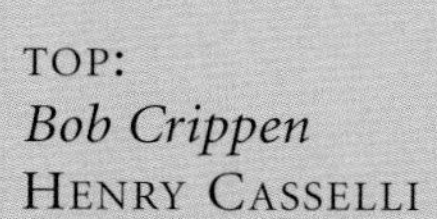

TOP:
Bob Crippen
HENRY CASSELLI

ABOVE:
John Young
HENRY CASSELLI

ABOVE RIGHT:
Preparations of the Space Shuttle Columbia, April, 1981,
FRANKLIN MCMAHON

RIGHT:
1:07:22 and Counting,
ARTHUR SHILSTONE
Scores of media reporters and cameras record Columbia's assent into space.

COLUMBIA'S FIRST FLIGHT

As John Young and I lay on our backs on that morning of April 12, 1981, confined by our bulky escape suits and tightly strapped into Columbia's ejection seats, it almost seemed as though we were performing another simulation. We'd gone through this exercise many times since being selected three years earlier to fly the initial test flight of the space shuttle. In fact, we'd been here only two days earlier, having spent six hours confined in the cockpit before the launch was finally scrubbed due to a computer problem.

Yet here we were, once again. Although the shuttle team had done a marvelous job of solving the problem, I truly thought we would end up in another scrub. This was a very complex vehicle—the world's greatest all-electric flying machine, as John liked to call it. It had not undergone unmanned flight testing and we would be the first humans to fly in it. I was lucky to be flying with someone like John, a veteran of four previous flights, including both Gemini and Apollo. Columbia was to be my first flight, even though I had been an astronaut with NASA for twelve years.

John had said that if they were going to light off seven million pounds of thrust under you and you weren't a little bit excited, you didn't understand the situation. Both John and I understood the situation. We had worked on the design of the shuttle from its inception. Still, the excitement didn't really hit me until the countdown passed one minute to go. It was then that I turned to John and said that I thought we really might do it. The whole shuttle team had worked years for this and here was the moment of truth. Launch control recorded my heart rate at that time at over 120 beats per minute. I'm surprised it wasn't higher.

With the hydraulic systems pumping life through her veins, Columbia felt alive. At minus six seconds, the main engines ignited and the cockpit became noisy as the vehicle rocked forward. Once we returned to vertical, the big moment arrived. The solid rocket motors ignited and we were off, no doubt about it. The sensation reminded me of the aircraft-carrier catapult shots I had experienced early in my Navy flying career. Shortly after clearing the launch pad, the vehicle went through its planned roll program to align itself for our launch direction. The roll caused concern among many of the spectators since, unlike on symmetrical launch vehicles like the early Saturn, it was much more visible on a winged vehicle.

First stage was quite a ride. The solid rockets were quite loud and produced a pronounced low-frequency vibration. (We later learned that their ignition had reflected off the launch pad and actually damaged some of the structural components in the nose of the vehicle. In subsequent flights, troughs of water were placed in the launch-pad flame bucket.) Both John and I observed the trajectory begin to loft above what was planned. Although not so great as to be a serious concern, it was obvious that the vehicle was producing more lift than our engineering data had predicted. Two minutes later, the solids expended their fuel and separated from the vehicle with a fiery blast from the jettison rockets.

Then came my first moment of alarm. It seemed too quiet. My trust level had changed dramatically—I thought our main engines had quit, but a quick check of the instruments verified that everything was fine. We were just so high that no atmosphere existed to allow engine noise to reach the cockpit. Columbia once again began accelerating as the engines consumed the hydrogen and oxygen from the external tank connected to her belly. Upon reaching three g's (three times the normal acceleration of gravity), the engines automatically began to throttle to keep the acceleration constant.

Most of my attention throughout the flight had been focused on my instruments and computer displays. I watched the speed continue to build until, at 17,500 miles per hour and eight and a half minutes after liftoff, it was over. The main engines shut down and we were in orbit. It was the fastest eight and one half minutes I have ever experienced.

It was then, finally, that I had time to glance out of the window and observe Earth from space—one of the most beautiful and dramatic sights I've ever beheld. I radioed back to my colleagues, who had also been waiting many years for this event, and told them it was well worth it.

BOB CRIPPEN

ELECTROHOME
ELECTROHOME
:32
:30
B.M.JACKSON '81

Time, Space, and Columbia
BILLY MORROW JACKSON
The Columbia STS-2 mission still managed to attract many reporters. Jackson can be seen in the foreground wearing a black cap.

TOP:
Sixty Seconds to Touchdown
Nixon Galloway
A gigantic sonic boom marks Columbia's reentry into the Earth's atmosphere.

ABOVE:
After Images
Karen Chandler
Shuttle convoy teams perform post-flight operations immediately following the shuttle's return to Earth.

OPPOSITE:
Mission STS-4: Challenger, Edwards Air Force Base
Ron Cobb
The orbiter Challenger sits atop a Boeing 747 prior to transportation to Kennedy Space Center.

One on the Pad, One on the Bay
Joe Jaqua
The Space Shuttle Atlantis, visible on monitor at bottom left, waits for launch while workers in the Vehicle Assembly Building prepare Space Shuttle Discovery for mission STS-33.

PAGES 134–35:
Shuttle Night Power
Tom Newsom
A Challenger liftoff.

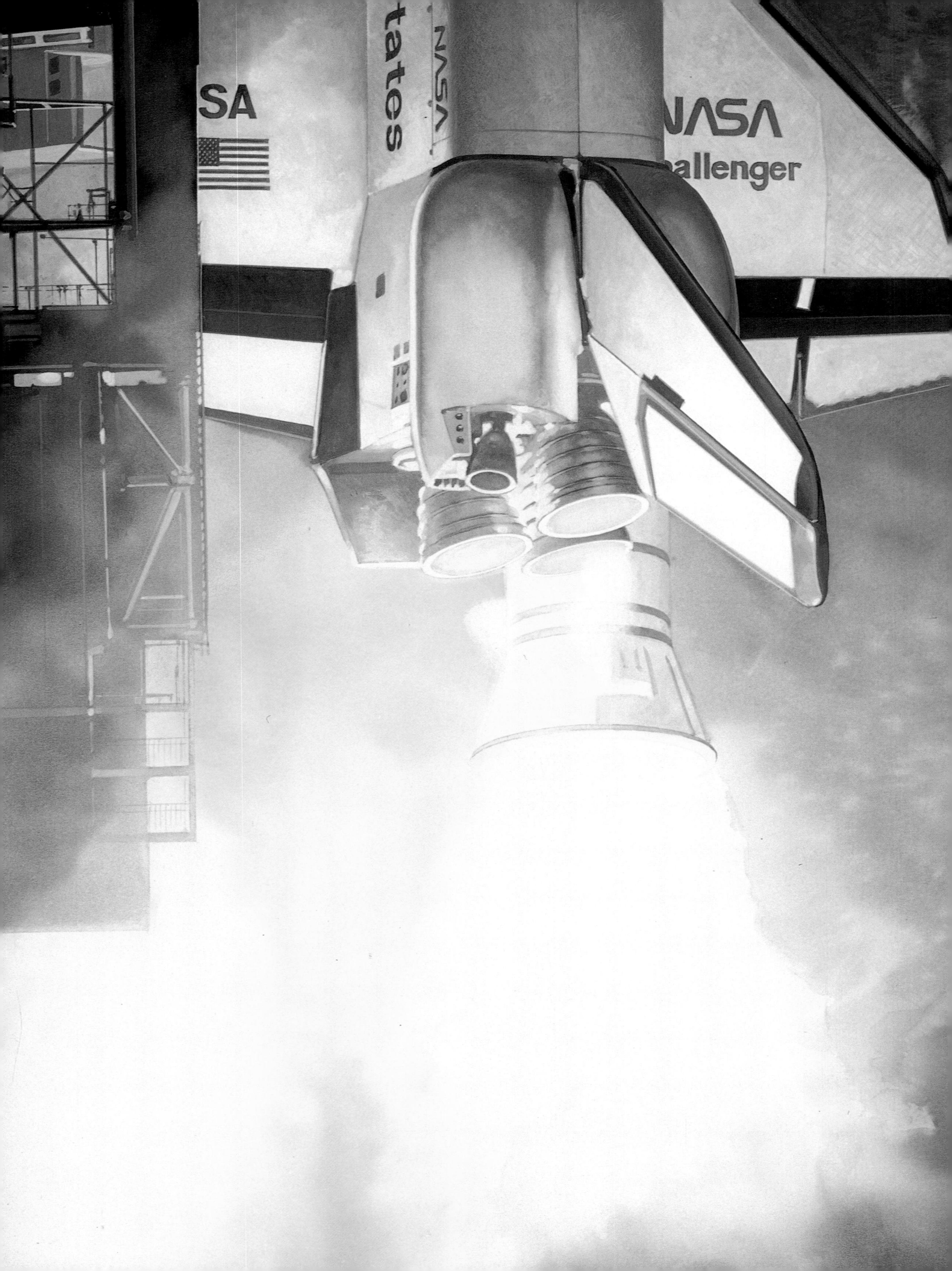

SA
tates
NASA
NASA
allenger

The Tile Team, Morton Kunstler
Replacing tiles on the belly of the shuttle orbiter.

ORBITER PROCESSING FACILITY (OPF)

When birds aren't flying, they're generally in the nest. And such is the case with NASA's four shuttle orbiters, or "birds," as their technicians like to call them. For these birds, the nests are the three Orbiter Processing Facility (OPF) hangers, located to the north and west of the giant Vehicle Assembly Building (VAB) and where the orbiter fleet spends most of its time before and after a mission into low earth orbit.

The OPF plays a critical role in processing for flight the four existing orbiters in NASA's shuttle fleet: Columbia, Discovery, Atlantis, and Endeavour. (Endeavour is the replacement vehicle for the Challenger, which was lost in a January 1986 launch.) OPF 1 and 2 were originally built in the late 1970's; OPF 3 was constructed following the reflight period after the Challenger accident. At 95 feet high, 397 feet long, and 233 feet wide, these facilities offer plenty of room to service an orbiter (The Shuttle is referred to as an orbiter when not attached to its external tank or solid rocket boosters). An OPF features multiple platform levels and walkways that allow engineers and technicians access to every inch of the spacecraft. Since each shuttle is expected to fly at least 100 missions, comprehensive access allows for meticulous attention to be paid to the thousands of components in order to ensure that a Shuttle will launch safely, withstand space temperatures, and glide home intact.

After a flight, an orbiter is housed inside the environmentally controlled OPF, where its main engines are removed and replaced. Payloads are emptied and the payload bay is reconfigured for the next mission. The orbiter's landing gear is serviced and the tires are replaced. While these and hundreds of other functions are taking place, each of the approximately 24,000 individual thermal protection system tiles and the 2,300 flexible insulated quilts are inspected and, if necessary, replaced.

Orbiter tiles are made from low-density, high-purity silica (99.8 percent) amorphous fiber (fibers derived from common sand) insulation made rigid by ceramic bonding. Individual tile thickness varies from one to five inches; a single tile is about the size of a paperback book and weighs a mere ounce, yet provides the orbiter with critical protection during re-entry against temperatures as high as 2300°F. The flexible quilts adhere to the sides and top of the orbiter where the least amount of re-entry heat is felt, while the reusable surface insulation tiles cover the entire underside of the vehicle. Though resembling a giant jigsaw puzzle, the tiles are actually individually encoded for identification and placement.

Tiling the Space Shuttle Columbia
Charles Schmidt
OPF technician adheres a shuttle tile.

The various functions, verifications, and tests performed on the orbiter's multiple systems range from special tests of systems that are flown repeatedly and require only routine servicing to systems that necessitate replacement, such as the hypergolic propellant system, which must be completely drained of residual hazardous propellants and refilled prior to each mission. Also, explosive ordnance items—used to deploy the drag chute and assist in the deployment of the main landing gear and nose landing gear, if necessary—are removed after each flight and checked before each successive mission. Such servicing can be safely completed in only the Orbiter Processing Facility.

Overall, turnaround time for servicing has been reduced tremendously since the first space-shuttle launch took place. At that time, Columbia, spent 610 days in the OPF prior to launch. Today, the time an orbiter spends in the OPF is usually 70 to 80 days, after which it is taken to the Vehicle Assembly Building, mated to the external tank and twin solid rocket boosters, and sent into space. When its mission is completed, the orbiter undergoes the entire OPF process once again.

Bruce Buckingham, Kennedy Space Center

MISSION STS-51L

First Teacher in Space
Robert L. Shaar
Christa McAuliffe, the first teacher in space, was to instruct youngsters from across the country during the mission.

BELOW:
Hail to the Long Distance Voyagers, In Loving Memory
Robert A. M. Stephens
A drifting cloud over the empty launch pad symbolizes the reflective and reverent moments following the Challenger tragedy.

rocket boosters that ignited the main liquid-fuel tank. The televised explosion became one of the most significant events of the 1980s. Countless millions around the world grieved for the loss of the Challenger's seven crew members—Dick Scobee, Ron McNair, Mike Smith, Judy Resnik, Ellison Onizuka, Greg Jarvis, and civilian teacher Christa McAuliffe—who represented a cross-section of the American population in terms of race, gender, geography, background, and religion.

Following the Challenger accident, several investigations took place to determine the cause of the tragedy and to ascertain what changes should be made to the program to ensure shuttle safety and reliability. NASA added a drag-chute system and made extensive landing safety improvements, including upgrades of the orbiter fleet's tires, brakes, and nose-wheel steering mechanism. Engineers also made other numerous safety improvements, such as the installation of a crew escape system, which would allow astronauts to parachute from the orbiter under certain conditions. Finally, NASA completely reorganized shuttle-program management.

Since its resumption of flight operations with the launch of Discovery on September 29, 1988, the shuttle has returned to its former status as a workhorse of space exploration for both international and domestic projects. From its first flight in 1981 to the flight in March 1998, there have been 96 shuttle missions flown, including, since 1986, the Magellan spacecraft launch to Venus, the Galileo spacecraft to Jupiter, and the international Ulysses space-

MISSION STS-26

Star Kachina Spirit Orbiter
DAN NAMINGHA
The Star Kachina spirit floating with the shuttle reflects the Hopi-Tewa ancestry of the artist.

Imaging to the Edge of Space and Time
James L. Cunningham
The two concentric rings represent the epochs of the universe—the Big Bang and the present day; they also symbolize the Hubble Space Telescope's mirrors.

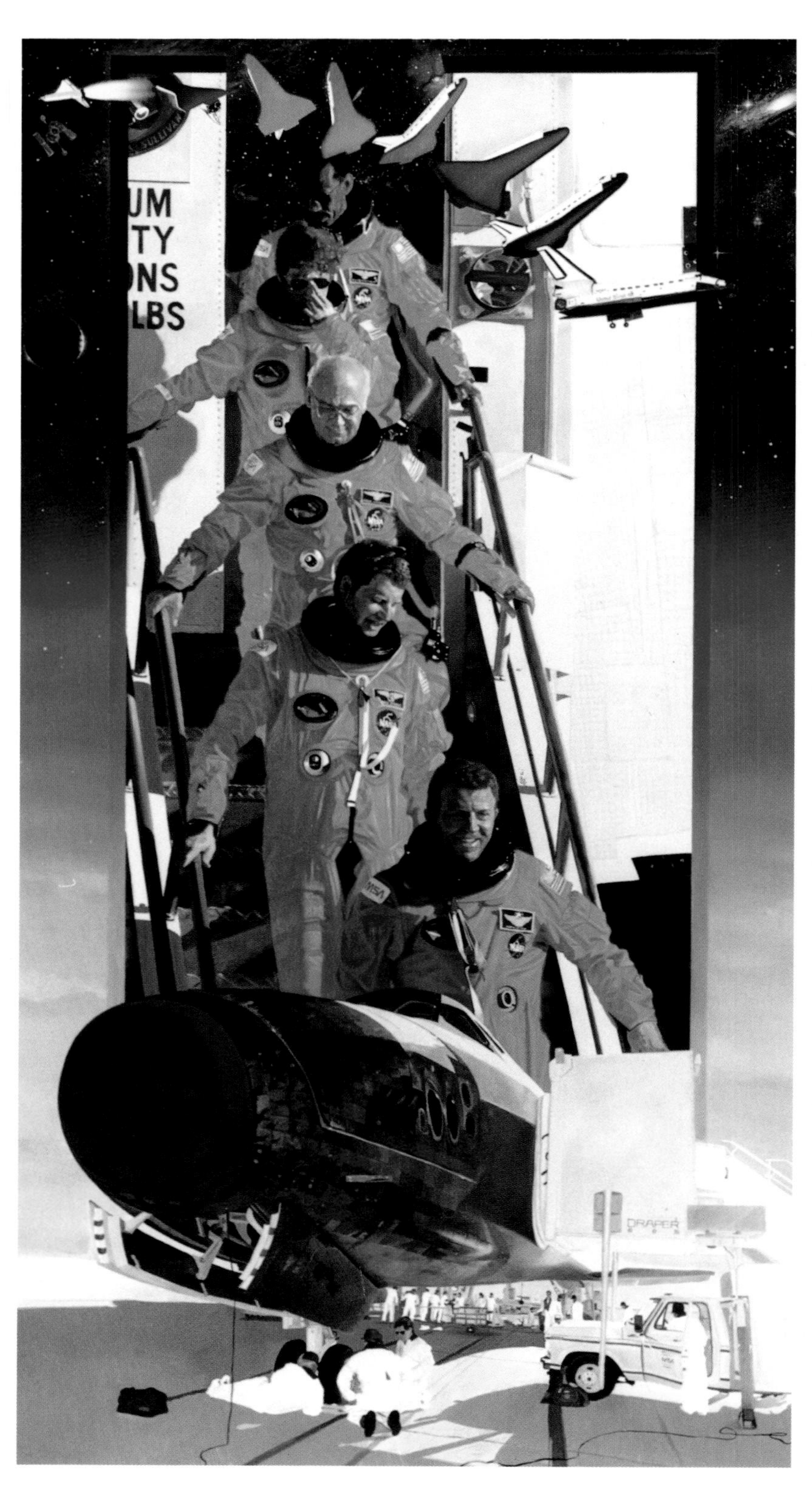

The Wind Beneath the Wings
Linda Draper
STS-31 astronauts exit the Space Shuttle Discovery.

LEFT:
Dry Lake Landing (Study)
P. A. Nisbet
Field studies of dry lake landing.

BELOW:
Dry Lake Landing STS-31
P. A. Nisbet
Discovery lands on the desert bed of Edwards Air Force Base on April 29, 1990, marking the return of the STS-31 crew.

craft to study the Sun. The shuttle also has deployed the Gamma Ray Observatory, the Hubble Space Telescope, and the Upper Atmosphere Research Satellite. Between April 1981 and March 1998, it carried approximately 2.3 million pounds of cargo and more than 750 major payloads into orbit, including more than 300 for NASA, 143 for the Department of Defense, 106 for commercial interests, 76 for other nations, and 74 educational payloads. Through 1997, astronaut crews have also conducted more than fifty extravehicular activities (EVAs), or spacewalks.

The space shuttle enjoys the same plaudits and suffers from the same criticisms that have been made clear since the program first began. It remains the only vehicle in the world with the dual capability to deliver and return large payloads to and from orbit. The design, now more than two decades

THE GAMMA RAY OBSERVATORY

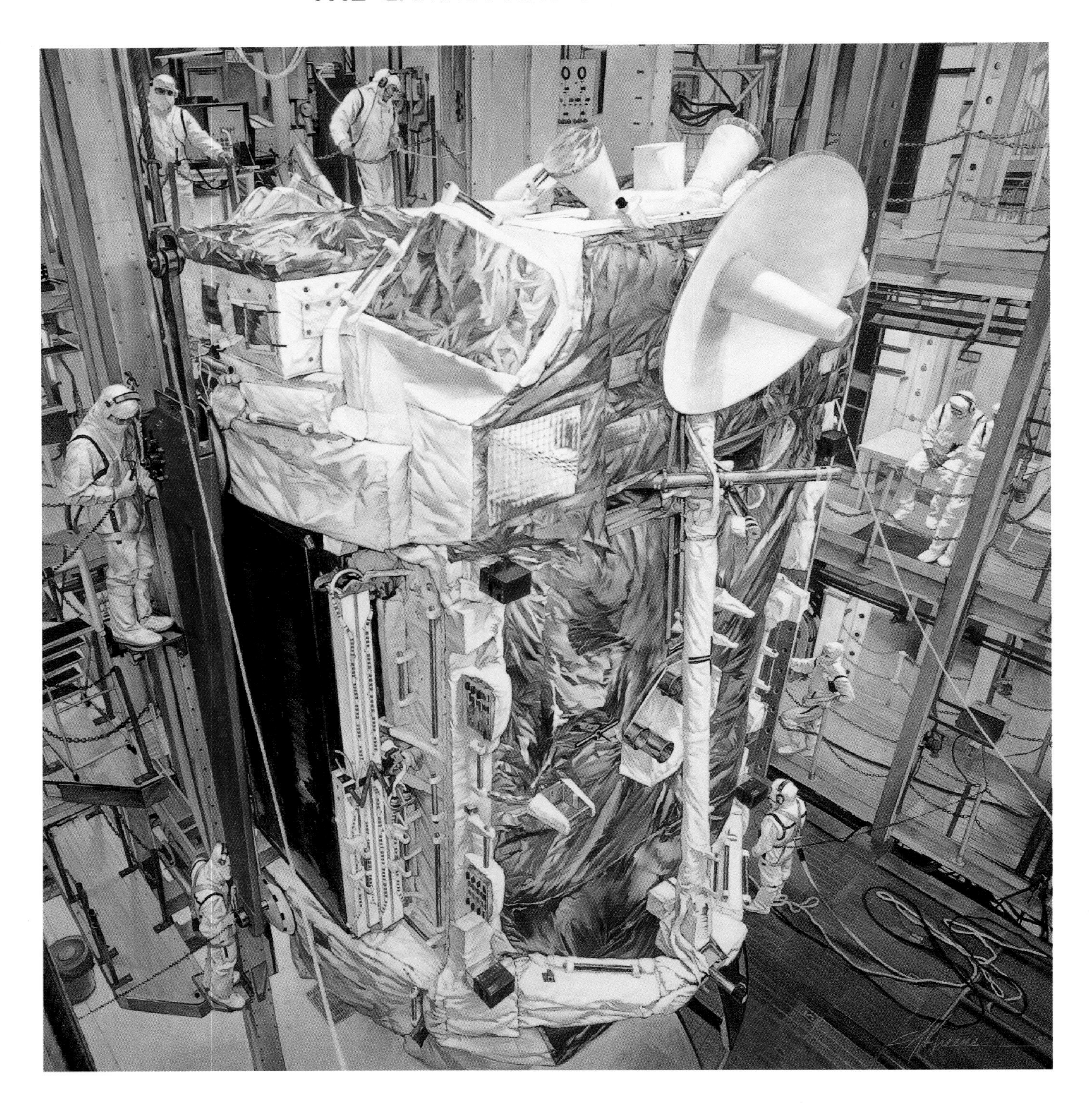

Preparing the Gamma Ray Observatory
NATHAN GREENE
The Gamma Ray Observatory is hoisted to a test cell in the Vertical Processing Facility at Kennedy Space Center.

GRO Deployment
NATHAN GREENE
Space Shuttle Atlantis deploys the
Gamma Ray Observatory, April 1991.

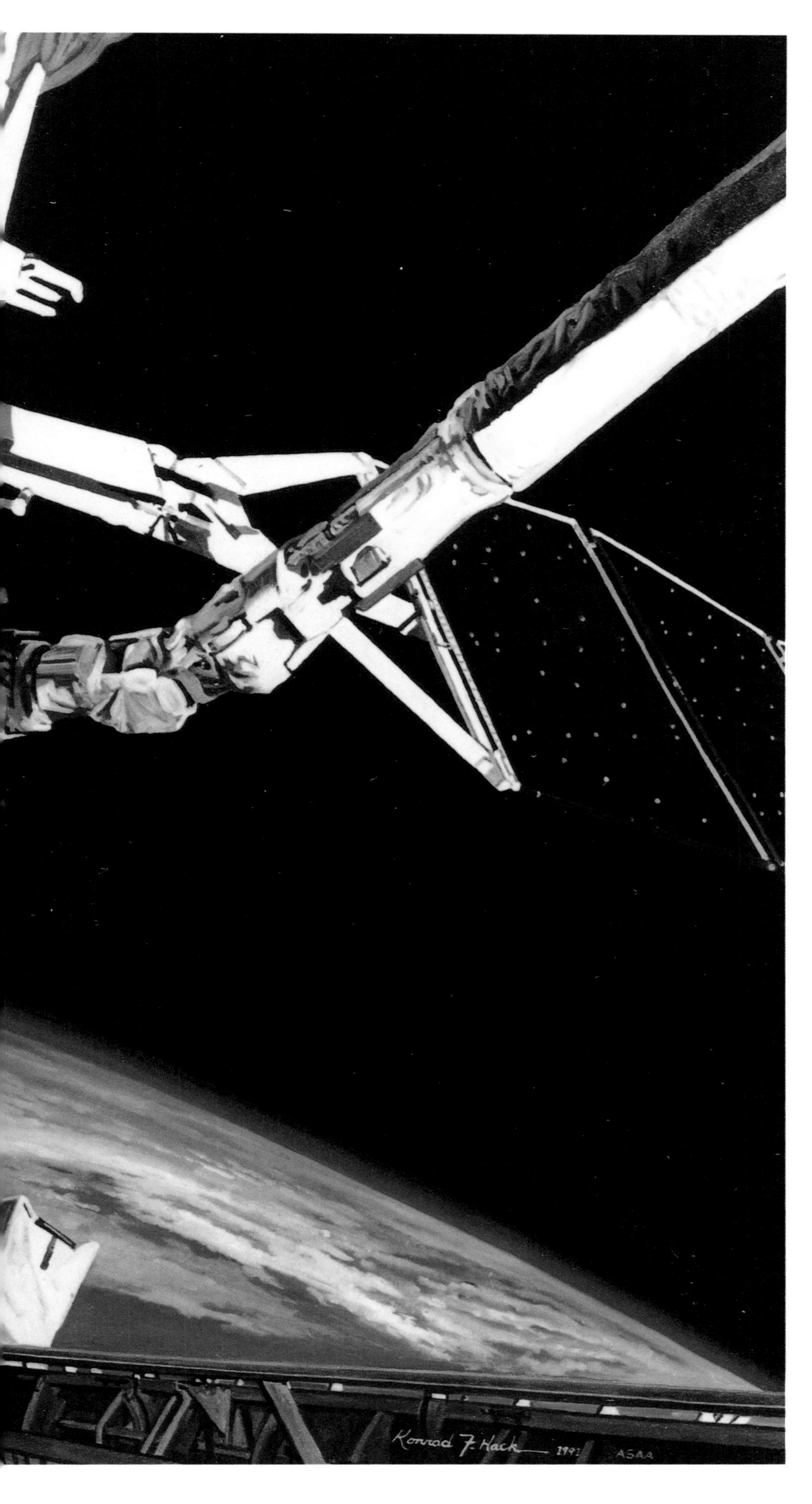

Deploying GRO
KONRAD HACK
Astronauts perform an EVA, or extravehicular activity, prior to deploying the Gamma Ray Observatory.

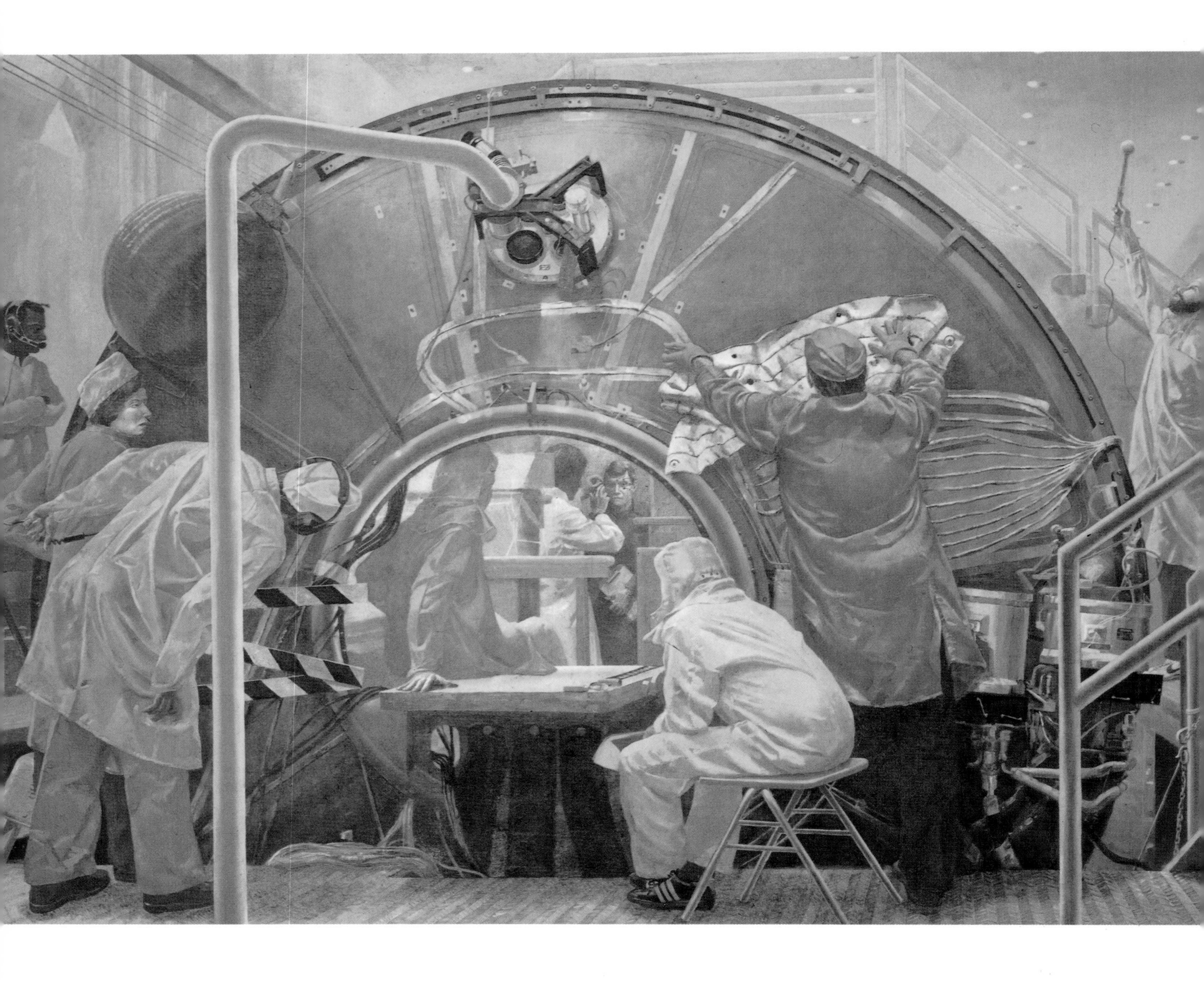

Space Lab 1, STS-9
CHARLES SCHMIDT
NASA and ESA (European Space Agency) astronauts and technicians work on the scientific laboratory prior to launch.

old, is still state-of-the-art in many areas, including computerized flight control, airframe design, electrical power systems, thermal protection system, and main engines. It is also the most reliable launch system currently in service anywhere in the world, with a success-to-failure ratio of better than ninety-eight percent.

As for international involvement, the space shuttle expanded by at least an order of magnitude the opportunity for human space flight, and many of the beneficiaries of this were foreign astronauts. During the years of the space shuttle's development and operations, from the 1970's into the late 1990's, international cooperation blossomed between the United States, Canada, and the nations of western Europe, whose rising economies were increasingly able to support the high costs of space science and technology. By the time of its tenth anniversary in 1991, the space shuttle had flown 204 people, some of them more than once. For example, between November and December 1983 on Columbia (STS–9), the first Spacelab mission took place and included the

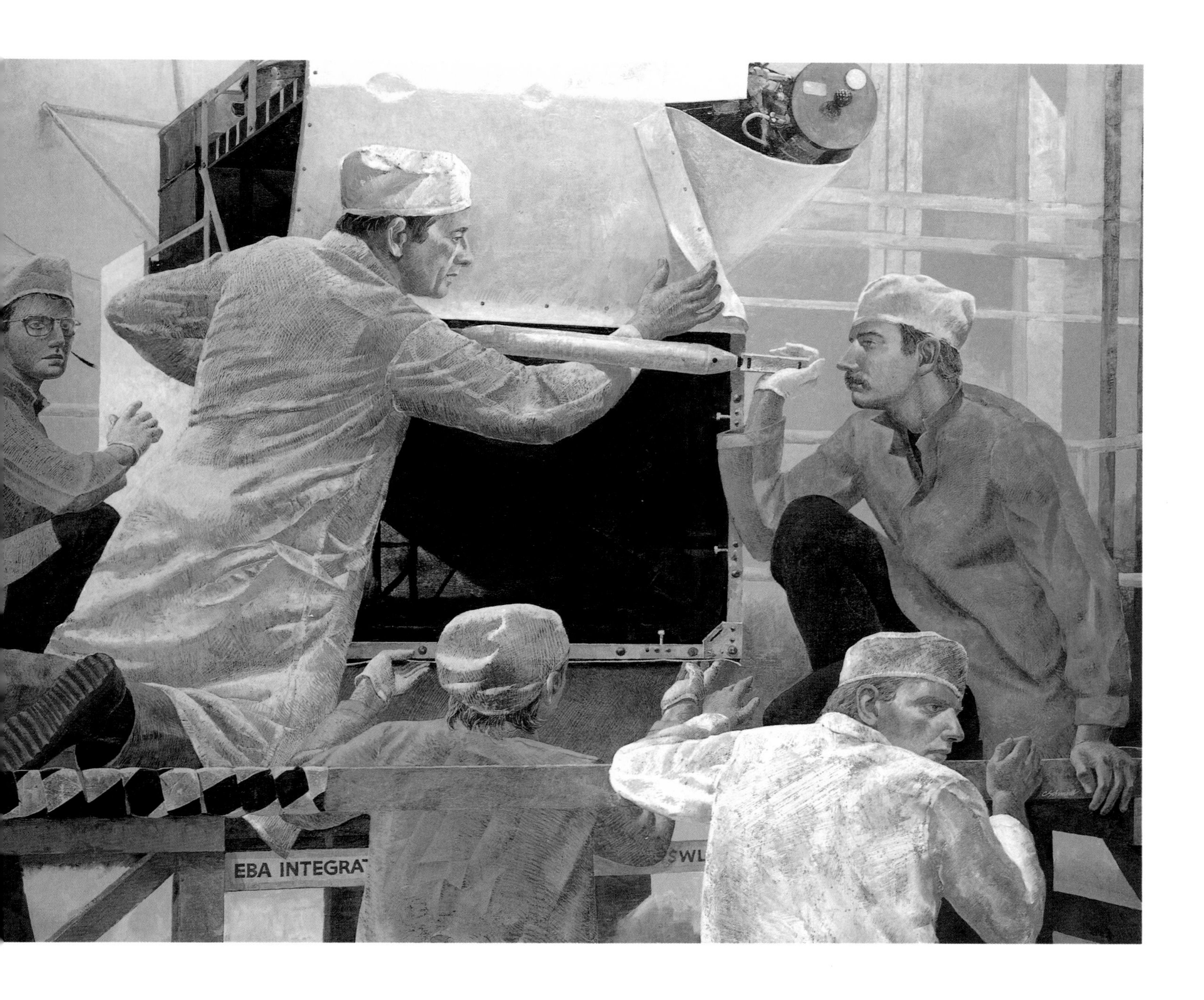

first non-U.S. astronaut, the West German Ulf Merbold. A year later, the first Canadian, Marc Garneau, flew Challenger (STS-41G). In June 1985 the first French and Saudi Arabian (Sultan bin Salman bin Abdul-Aziz Al-Saud) crew members flew on Discovery (STS-51G). A succession of foreign astronauts have flown on the shuttle since that time, many from member nations of the European Space Agency, Japan, Mexico, Canada, and post-1993 Russia.

Space Lab 1, Payload Integration
CHARLES SCHMIDT
Technicians perform tests to ensure a safely installed payload.

The cause of space science at the international level also received a significant boost from the shuttle since its capacity enabled large numbers of experiments to be carried out onboard. The shuttle's contributions to atmospheric and space physics, high energy astrophysics, infrared astronomy, optical and ultraviolet astronomy, life sciences, and materials research have been especially important. The shuttle also allowed the release of international payloads for all kinds of scientific and commercial purposes. One of the successes of the shuttle-science program was the "Get Away Special," a set of containers available on all shuttle missions that gave both professional and nonprofes-

TOP: *A Tear of Pride*
ABOVE: *From the Heavens*
WELLINGTON WARD

"As I stood on the grassy plot next to the quiet waters at the Kennedy Space Center, my full sensitivity as a human being savored the sights and sounds of the shuttle as it roared ever upward with majesty and grace. As I turned to observe the reactions of the faces in the crowd, I saw reflected in the sunglasses of a few, a drama that was unfolding before me. Although the glasses mirrored the action of the moment, it was the facial expressions of those watching that told the real story; one of honor, glory, patriotism. Just a single tear on the face of one summed up the beauty of the moment. It was a tear of pride."

ARTIST WELLINGTON WARD

sional experimenters from throughout the world access to space for small-scale experiments. Spacelab, of course, represented an especially sophisticated laboratory available to researchers. During its use, it completed a remarkable series of science missions that have fundamentally affected the understanding of microgravity, life sciences, and assorted physical sciences.

In spite of the undeniable success of the space shuttle, ranging from international politics to scientific return on investment to institutional survival, NASA's space-shuttle program had reached a crossroads by the 1990's. Obviously, the promise of "routine access to space" and inexpensive operations had not been realized. The space shuttle, although a successful program in many ways—especially as a demonstration of technological ability and an important stop on the route toward a truly viable spaceplane—was an enormous disappointment in many others. As the twentieth century came to an end, it was increasingly clear that a new launch vehicle had to be developed to enter the frontier of the twenty-first century.

THE SPACE STATION

Virtually from the beginning of the twentieth century, those interested in the human exploration of space have envisioned the building of a massive Earth-orbital space station which would serve as a jumping-off point to the Moon and the planets. A permanently occupied space station would be a necessary outpost in the new frontier of space. The more technically minded recognized that once humans had achieved Earth-orbit at about 200 miles up—the presumed location of any space station—atmosphere and gravity could be controled and persons would be halfway to their ultimate destination.

During the 1950's, Wernher von Braun argued for an integrated space plan that centered on human exploration of the solar system. It involved basic ingredients accomplished in this order: (1) an Earth-orbital space station serviced by a reusable space vehicle; (2) a lunar base; and (3) a human mission to Mars. Von Braun espoused these ideas in a series of important magazine articles in *Collier's* over a two-year period in the early 1950s, which featured striking images by some of the best artists of the era.

At the end of the 1960s NASA developed a very important initiative to apply Apollo technology against the longstanding dream of building a space station. Skylab, a 100-ton orbital workshop that could be tended by astronauts, was launched into orbit on May 14, 1973, by the final Saturn V launch vehicle. Almost immediately, however, technical problems developed due to liftoff vibrations. Sixty-three seconds after launch, the meteoroid shield—designed also to shade Skylab's workshop from the Sun's rays—ripped off, taking with it one of the spacecraft's two solar panels; a piece of the lost panel wrapped around the remaining one and kept it from properly deploying. NASA's mission-control personnel maneuvered Skylab so that its Apollo Telescope Mount (ATM) solar panels faced the Sun to provide as much electricity as possible. However, due to the loss of the meteoroid shield, this positioning caused workshop temperatures to rise to 126 degrees Fahrenheit.

In an effort to remedy the situation, astronauts Charles Conrad, Jr., Paul J. Weitz, and Joseph P. Kerwin lifted off from Kennedy Space Center on May 25, 1973, in an Apollo capsule atop a Saturn I-B and rendezvoused with the orbital workshop. The crew carried a parasol, tools, and replacement film to

Satellite in Space
Lonny Schiff
An abstract of a deployed space station and its docking area.

repair the orbital workshop. After substantial repairs requiring extravehicular activity (EVA), the crew successfully deployed a parasol sunshade that cooled the inside temperature to 75 degrees Fahrenheit, making the workshop habitable. During a second June 7 EVA, the crew also freed the jammed solar array and increased power to the workshop.

Despite the workshop's early and recurring mechanical difficulties, NASA nevertheless benefited from the scientific knowledge gained regarding long-duration space flight. NASA flew three Skylab missions for a total of 171 days and 13 hours—totaling more hours spent in space and in EVA maneuvers than the combined hours spent in all of the world's previous space flights.

At the conclusion of the final occupied phase of the Skylab mission, ground controllers positioned Skylab into a stable attitude and shut down its systems. It was expected that Skylab would remain in orbit eight to ten years, by which time NASA might be able to reactivate it. In the fall of 1977, however, the system entered a rapidly decaying orbit and reentered the Earth's atmosphere two years later. On July 11, 1979, it finally impacted with the Earth's surface; the resulting debris dispersion area stretched from the Southeastern Indian Ocean to across a sparsely populated section of Western Australia. It was an inauspicious ending to the first American space station, but it had opened some doors of understanding and whetted the appetite of NASA leaders for a full-fledged space station.

PAGES 152–153:
Space Station Freedom
VINCENT DI FATE
An earlier 1980's concept of a space station.

THE INTERNATIONAL SPACE STATION

With the shift to the space shuttle as a system in 1981, NASA returned to its quest for a real space station as a site of orbital research and a jumping-off point to the planets. NASA administrator James M. Beggs persuaded President Ronald Reagan to endorse the building of a permanently occupied space station. In 1984, Reagan declared that "America has always been greatest when we dared to be great. We can reach for greatness again. We can follow our dreams to distant stars, living and working in space for peaceful, economic, and scientific gain. Tonight I am directing NASA to develop a permanently manned space station and to do it within a decade."

From the outset, both the Reagan administration and NASA intended Space Station Freedom to be an international program. Although a range of international cooperative activities had been carried out in the past—Spacelab, ASTP, and scientific data exchange—the station offered an opportunity for a truly integrated effort. The inclusion of international partners, many now with their own rapidly developing space-flight capabilities, could enhance the effort.

As always, NASA leaders were leery regarding the sharing of U.S. technology and the possibility of having to compromise their control over the

Space Station Module,
NATHAN GREENE
Technicians and engineers at Marshall Space Flight Center work on space-station modules.

V. DI FATE '87

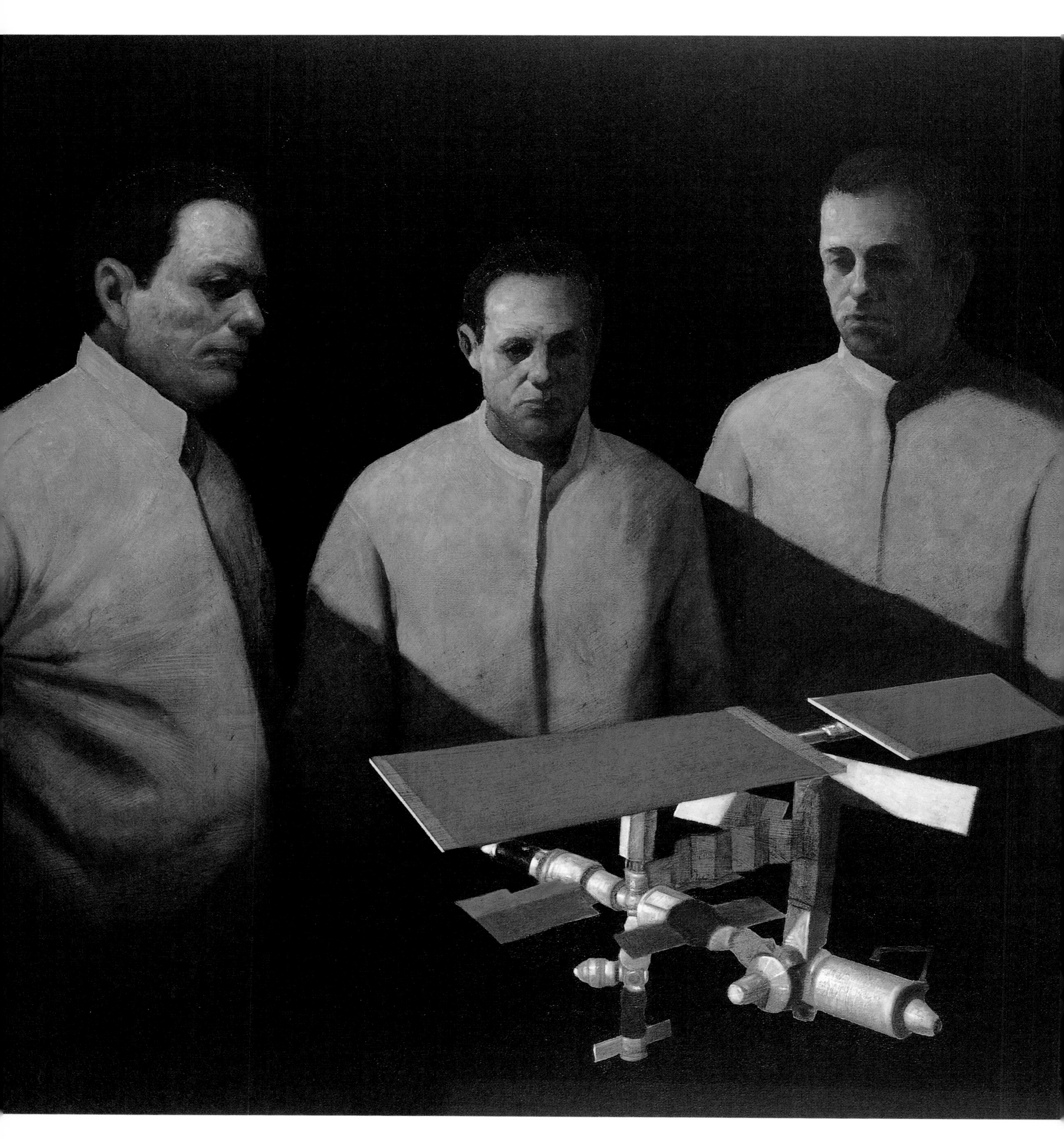

Dreams
CHRISTIAN VINCENT
Scientists and technicians work on a model of the International Space Station.

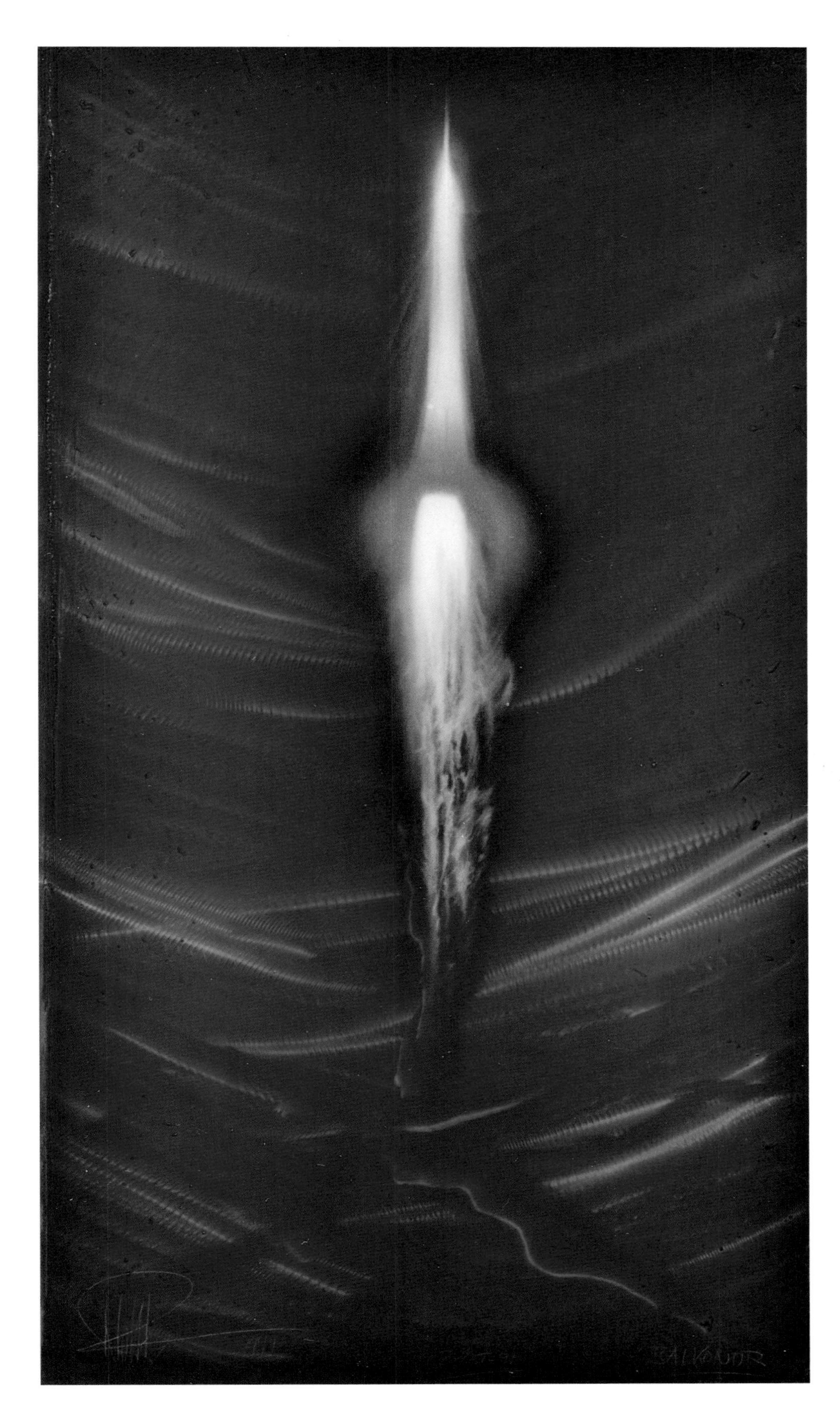

Soyuz Launch
ANDREAS NOTTEBOHM
A Soyuz rocket hoists Norm Thagard into space to begin his historic stay on the Russian space station Mir.

execution of the program. Despite these concerns, however, they pressed forward with international agreements allowing thirteen nations to take part in the Space Station Freedom program. Japan, Canada, and the nations pooling their resources in the European Space Agency (ESA) agreed in the spring of 1985 to participate. Canada, for instance, built a remote servicing system. Utilizing its Spacelab experience, the ESA built an attached pressurized science module and an astronaut-tended free-flyer. Japan contributed the development and commercial use of an experiment module for material processing, life sciences, and technological development. These separate components, with their "plug-in" capacity, somewhat eased the concerns of NASA management, as well as Congress, regarding unwanted technology transfer.

The Space Station Freedom program was nevertheless controversial. Debate centered on its costs versus its benefits. As a result, NASA designed the project to fit an $8 billion research-and-development funding profile. Within five years, the projected costs had more than tripled and the station had become too expensive to fund fully during the national-debt crises of the 1980s, necessitating annual redesigns of the space station between 1990 and 1993. The project became smaller and retained less and less of its original broad vision. As costs were reduced and capabilities diminished, political leaders who had once supported the program began questioning its viability. Some leaders suggested that the nation, NASA, and the overall space exploration effort would be better off if the space-station program were terminated. After a few years of additional study and planning, NASA came forward with a more promising vision.

NASA Administrator Daniel S. Goldin proposed to the government three redesign options for the space station, and on June 17, 1993, President Bill Clinton decided to proceed with a moderately priced, moderately capable station. At nearly the same time, the dissolution of the USSR allowed NASA to negotiate a landmark decision to include Russia in the building of an international space station. On November 7, 1993, a joint announcement was made by the United States and Russia that they would work together with twelve international partners to build a space station for the benefit of all. Even so, the fourteen-nation international space-station program remains a difficult issue, and public policymakers wrestle with competing political agendas lacking consensus.

Twenty years after the Apollo-Soyuz Test Project, the United States and Russia again met in Earth orbit when the Space Shuttle Atlantis docked with the Russian Mir space station in the summer of 1995. "This flight heralds a new era of friendship and cooperation between our two countries," said NASA Administrator Daniel S. Goldin. "It will lay the foundation for construction of an international space station later this decade."

The Atlantis mission was the first of nine planned shuttle-Mir linkups between 1995 and 1997, including rendezvous, docking, and crew transfers. These flights were intended to pave the way toward the assembly of an international space station set to be constructed in orbit beginning in late 1998. Atlantis lifted off on June 27, 1995, from Kennedy Space Center and docked with Mir on July 29. After ceremonies following the rendezvous and docking, the two groups of spacefarers undertook several days of joint scientific investigations inside the Spacelab module tucked in Atlantis's large cargo bay. On July 4, 1995, following joint activities, two cosmonauts and astronaut Norman

OPPOSITE:
Cosmonaut Vladimir N. Dezhurov, EVA at Mir
John Clendening
Dezhurov performs EVA activities during the Mir docking mission of 1995.

STS-71

Edgar H. Sorrells-Adewale

The Mir docking mission celebrated friendly cooperation between Russians and Americans.

H. Thagard, all of whom had been aboard the station since March 16, 1995, joined the shuttle crew for a return trip to Earth. Thagard returned home with the American record for a single space flight with more than 100 days in space. The Atlantis returned to the Kennedy Space Center on July 7.

Although there is still much to accomplish, the docking missions conducted through 1997, which were aimed toward increasing international space-flight capabilities, seemed also to signal a major alteration in the history of space exploration. With the launch of the first international space-station components in 1998, international competition has indeed been replaced with cooperation as the primary reason behind making huge space-operation expenditures possible. As the dean of space policy analysts, John M. Logsdon, concluded, "There is little doubt, then, that there will be an international space station, barring major catastrophes like another shuttle accident or the rise to power of a Russian government opposed to cooperation with the West. . . . and that for the next seven years the fourteen-state station partnership can focus all of its energies on finally putting together the orbital facility, without the diversion of continuing political arguments over its basic existence and overall character.

"Even with all its difficulties and compromises, the space station partnership still stands as the most likely model for future human activities in space. The complex multilateral mechanisms for managing station operations and utilization will become a de facto world space agency for human space flight operations, and planning for future missions beyond Earth orbit is most likely to occur within the political framework of the station partnership."

People several hundred years hence may well look back on the building of an international space station as the tangible evidence of the beginning of a cooperative effort that succeeded in creating a permanent presence for humans beyond the Earth. It could, however, prove to be only a minor respite in the competition between nations for economic, political, and technological supremacy.

CHAPTER FOUR

TOWARD FUTURE EXPLORATION

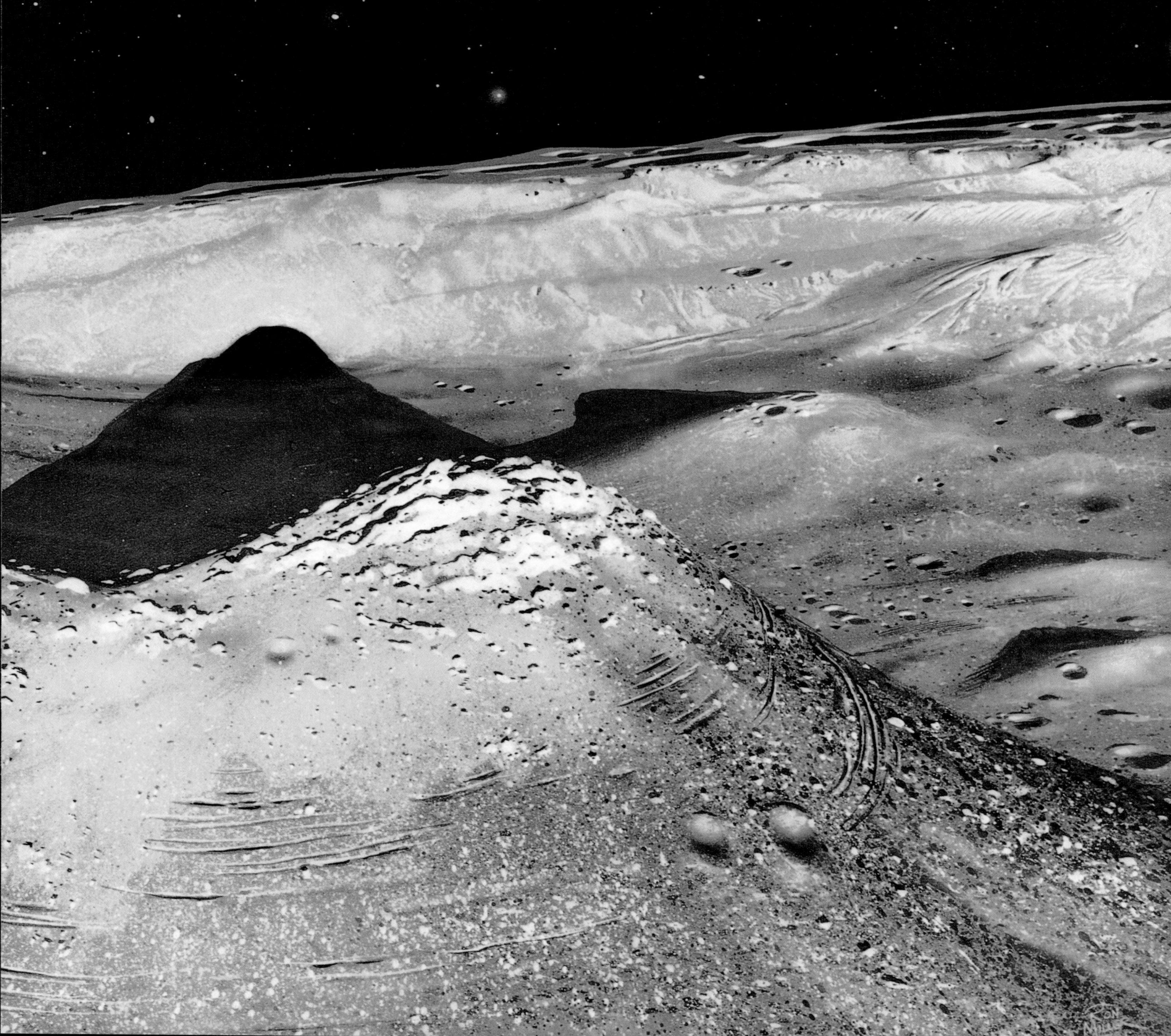

Space exploration today has the ability to excite and inspire modern society just as the exploration of the world beyond Europe held the fascination of fifteenth- and sixteenth-century societies. The discovery of a vast unknown and its potential capability of supporting human life holds a tremendous allure and is particularly appealing to a culture such as that of the United States, which has been heavily influenced by territorial expansion. Space exploration is, indeed, the "final frontier," and the opening up of this frontier can be a reality for all humanity in the twenty-first century. To accomplish it, however, requires the collection and assimilation of scientific and technological knowledge. This process, well underway, can trace its roots back four decades to the early space-science missions of the newly born NASA which, at that time, set forth a plan to progress in three distinct phases: (1) lunar exploration, (2) inner-planetary probes, and (3) missions to the outer planets.

Speck of Dust
PIERRE MION
Explorer astronauts are dwarfed by the Moon's vast surface.

OPPOSITE:
Black Holes Make Me Think of Light
CHARLES ROSS

PAGES 160–61:
View from Mimas
RON MILLER
A view of Saturn from its moon Mimas.

LUNAR EXPLORATION

A scant quarter of a million miles away in the context of cosmic distances, the Moon was an attractive early target for exploration and in 1958 the United States began a program to reach it. What emerged was Project Ranger, a program that ran between 1961 and 1962, which unfortunately suffered from repeated failures to achieve a correct orbit. After NASA engineers eliminated all of the Ranger's scientific instruments save for a television camera, the agency reorganized the project, assigning it one sole objective: to go out in a blaze of glory by taking high-resolution pictures before crashing into the Moon. On July 31, 1964, the seventh Ranger succeeded and transmitted 4,316 high-resolution pictures of the Moon's Sea of Clouds.

Two more robotic projects followed in quick succession, with Lunar Orbiter and Surveyor each paving the way for the Apollo landing efforts of the late 1960s. The Orbiter mapped Apollo landing sites from above the surface. Surveyor 1 landed successfully on June 2, 1966, and transmitted more than 10,000 high-quality photographs of broad panoramic views of the craters, hills, and rocks on the ancient lunar surface. By the completion of these programs, much additional scientific data had been gathered, both for the use of Apollo planners and for the broader lunar science community.

It took nearly a generation after the final Apollo flight until the United States would return to the Moon with the Clementine space probe, a joint project between the Department of Defense's Strategic Defense Initiative Organization and NASA. Operating on a shoestring budget, the principal mission of the Clementine spacecraft was to test sensors and spacecraft components under extended exposure to the space environment; the survey of the Moon was a by-product of this mission.

Launched on January 25, 1994, and taking a circuitous route, Clementine achieved lunar orbit on February 21, and for the next three months it photographed and mapped more than ninety percent of the lunar surface, adding extensive data to the scientific store. Using this early information, scientists in late 1996 determined that ice from an asteroid crash may exist on the Moon's south pole. Thus, lunar science has once again been energized and attention has been placed on missions to explore the area further. In January 1998, NASA launched the Lunar Prospector, which two months later confirmed the existence of ice at the Moon's southern polar region.

INNER-PLANETARY PROBES

In addition to the lunar missions, there have been significant scientific probes sent to the inner planets of the solar system. Two early targets, Venus and Mars, were chosen because of their proximity to Earth and the possibility that life might exist on them.

As the evening and morning star, Venus has long enchanted humans. Shrouded in a mysterious cloak of clouds that permanently hides its surface

The Firefly
CHARLES SCHMIDT
Mariner 9's historic fly-by of Mars in 1972 uncovered extraordinary geology, including the great valley Vallis Marinis.

from view, Venus is the closest planet to Earth and a near twin in terms of size, mass, and gravitation. These attributes have contributed to the many speculations and popular conceptions about the nature of Venus and the possibility of life existing there. In fact, one notion of the early twentieth century held that the Sun had been gradually cooling for millennia, and that as it did so, each planet in the solar system took its turn to be a haven for life. The possibility of a lush, watery Venusian world populated by reptilian creatures permeated the popular culture.

The United States claimed the first success in Venusian exploration when Mariner 2 flew by the planet in December 1962 at a distance of 21,641 miles. It probed the clouds, estimated planetary temperatures, measured the charged-particle environment, and looked for a magnetic field similar to Earth's magnetosphere, but found none. It then sped inside the orbit of Venus and eventually ceased operations in 1963. The United States sent another probe to Venus in 1975 when Mariner 10 traveled past the planet en route to Mercury. Two additional probes, Pioneer 12 and 13, were sent out in 1978,

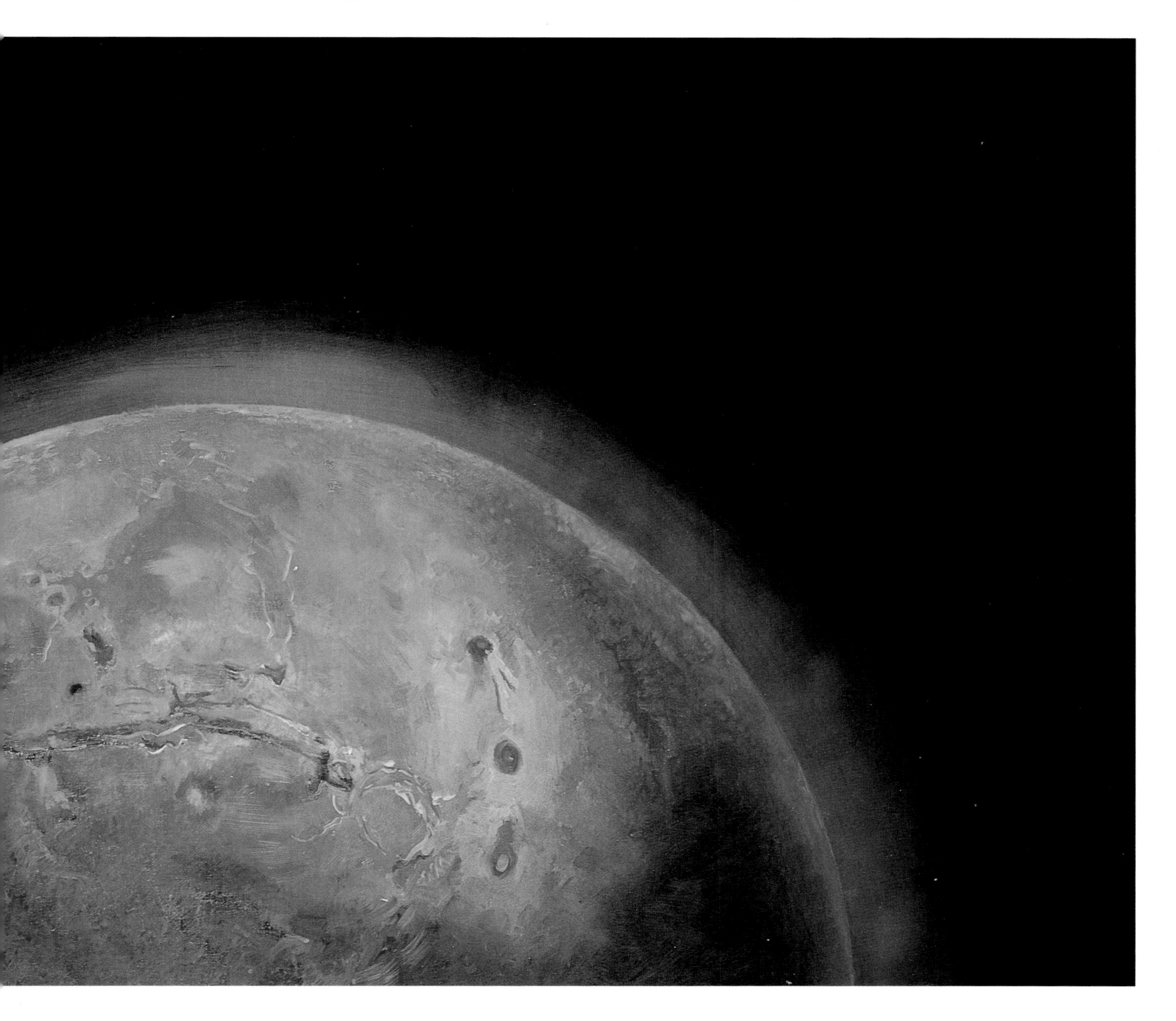

with Pioneer 13 releasing four heat-resisting probes into the planet's clouds to measure atmospheric pressure, winds, and temperature. Collectively, these probes revealed that Venus was superheated to about 900 degrees Fahrenheit due to the greenhouse effect of its cloud layer, and that its surface pressure was about ninety atmospheres—far greater than even the depths of the Earth's oceans. This data served the conclusion that the probability of life, as humans understood it, did not exist on Venus.

The best chance to map the features of the Venusian surface came in the early 1990s with a highly successful Magellan mission launched in 1989. Arriving at Venus in September 1990, Magellan mapped ninety-nine percent of the planet's surface at high resolution, with parts in stereo. The amount of digital imaging data returned was more than twice the sum of all returns from previous missions and provided some surprises. Among them was the discovery that plate tectonics were at work and that lava flows clearly showed evidence of volcanic activity. The conclusion of Magellan's mission in 1993 leaves years of research that awaits further analysis but has already fundamentally altered most of the theories of Venus as being a tropical, protoorganic planet.

Like Venus, Mars has long held a special fascination for humans as a planet on which life might exist. For a long time, its topographical features, known as canals, fed the popular imagination that Mars was once a watery planet and that the canals had been built by intelligent beings. It was only with the scientific data returned from the planet that this conception began to change, beginning with the Mariner probes.

On July 15, 1965, NASA's Mariner 4 flew within 6,118 miles of the planet and took twenty-one close-up images. Depicting a cratered, lunarlike surface without structures and canals or anything that remotely resembled a pattern produced by intelligent life, these photographs dashed the hopes of many that life resided on Mars. Between 1969 and 1971, NASA sent out Mariners 6, 7, and 9—probes that studied the planet's atmosphere and surface in order to lay the groundwork for what was later an enormously important Mars landing mission completed under the project name Viking.

It was the search for signs of life that prompted the Viking landing on Mars, and the excitement of possibly finding any was evident in the comments made by NASA Administrator James C. Fletcher on the eve of its launch in 1975: "Although the discoveries we shall make on our neighboring worlds will revolutionize our knowledge of the universe, and probably transform human society, it is unlikely that we will find intelligent life on the other planets of our Sun. Yet, it is likely we would find it among the stars of the galaxy, and that is reason enough to initiate the quest . . ." He then concluded: "It is hard to imagine anything more important than making contact with another intelligent race. It could be the most significant achievement of this millennium, perhaps the key to our survival as a species."

The Viking mission involved two identical spacecraft consisting of orbiters and landers. Viking 1 was launched on August 20, 1975, from Kennedy Space Center and landed on the Chryse Planitia (Golden Plains) on July 20, 1976. Viking 2 was launched for Mars on November 9, 1975 and landed on September 3, 1976. One of the most important scientific activities of this effort involved ascertaining whether life had ever existed on Mars, but the scientific data gained argued against the possibility, and after several years of contact,

Viking White Room
MARIO COOPER
Technicians and scientists prepare the Viking Lander.

and with a stream of data still providing grist for scientific debate, NASA finally closed down the landers in 1983.

Ironically, twenty years later, Viking's scientific findings have once again raised the possibility that life once existed on Mars. On August 7, 1996, a nine-member team of NASA and Stanford University scientists, led by Johnson Space Center scientists David S. McKay and Everett K. Gibson, Jr., announced they had uncovered evidence—not conclusive proof—of the one-time existence of microscopic life. The history of a four-pound meteor found in Antarctica in 1984 led them to suspect it was from Mars. Identified as ALH84001, the meteor was formed as an igneous rock around four and a half billion years ago when Mars was much warmer and probably contained oceans hospitable to life. When a large asteroid hit the planet nearly fifteen million years ago, it jettisoned the rock into space, where it remained until it crashed into the southern continent around 11000 BC. The team presented three compelling pieces of evidence that suggested fossillike remains of Martian microorganisms, dating back over three and a half billion years, are

present in ALH84001. During their two-and-a-half year investigation, the team found trace minerals in the meteor usually associated with microscopic organisms. A newly developed electron microscope uncovered possible microfossils that measured between 1/100 to 1/1000 the diameter of a human hair. They also discovered organic molecules called polycyclic aromatic hydrocarbons (PAHs), which usually result when microorganisms die and their complex organic molecules break down. Following these observations, the team called for additional research from other scientists that would either confirm or refute these findings about Mars.

Such research was, indeed, already underway, but it was not directly aimed at studying microbial Martian life. On July 4, 1997, the world watched in wonder as the inexpensive Mars Pathfinder, costing only $267 million, landed on Mars after its launch the previous December. A small, twenty-three-pound robotic rover named Sojourner departed from the main lander to record weather patterns, atmospheric opacity, and the chemical composition of rocks washed down into the Ares Vallis flood plain, an ancient outflow channel in Mars's northern hemisphere. Sojourner completed its milestone thirty-day mission on August 3, 1997, capturing far more data on the atmosphere, weather, and geology of Mars than scientists had expected. By the end of its mission in late 1997, Pathfinder had returned more than 1.2 gigabits (1.2 billion bits) of data and over 10,000 tantalizing pictures of the Martian landscape.

The geological history of Mars was further revealed with the Mars Global Surveyor, launched in December 1996. Soon after it entered the planet's orbit on September 11, 1997, the spacecraft's magnetometer detected the

LEFT:
Mars Rising
CHAKAIA BOOKER
The Pathfinder mission proved successful when the rover Sojourner (named after Sojourner Truth) explored the Martian surface in 1997.

OPPOSITE:
Viking/Mars Encounter
WILSON HURLEY
The Viking spacecraft after separation of the Lander from the Orbiter.

PAGES 170–71:
Tribute to Voyager
SARA LARKIN
Voyager reveals the outer planets of the universe.

existence of a planetary magnetic field. This held important implications for the possible development and continued existence of life there. Planets like Earth, Jupiter, and Saturn generate their magnetic fields by means of a dynamo made up of moving molten metal at their core. This metal is a very good conductor of electricity, and the rotation of the planet creates electrical currents deep within the planet that give rise to the magnetic field. A molten interior suggests the existence of internal heat sources, which could give rise to volcanoes and a flowing crust responsible for moving continents over geologic time periods. The data returned from this mission electrified both the scientific and public sectors, adding support for an aggressive set of missions to Mars by the year 2000 to help unlock the mysteries of the Red Planet and to point out the direction for future exploration and eventual colonization.

MISSIONS TO THE OUTER PLANETS

As Martian exploration was taking place, NASA also sent spacecraft to the outer Jovian planets of the solar system. In the early 1970's, Pioneers 10 and 11 paved the way with fly-bys of Jupiter and Saturn, but it was Voyager 1 and 2 that reached a high-water mark during the latter 1970's and early 1980's, a result of the discovery by scientists in the 1960's that the Earth and all the giant planets of the solar system aligned along one side of the Sun once every 176 years. This geometric lineup made possible the close-up observation of all the planets in the outer solar system (with the exception of Pluto) in a single flight, the "Grand Tour." As the spacecraft flew by each planet, its path would bend, thus increasing velocity sufficiently to deliver it to its next destination, a complicated "slingshot" effect known as "gravity assist." This process could reduce the flight time to Neptune from thirty to twelve years. To take advantage of this rare opportunity, NASA launched the Voyager missions from Cape Canaveral, Florida, with Voyager 2 lifting off on August 20, 1977, and Voyager 1 entering space on a faster, shorter trajectory on September 5, 1977.

OPPOSITE:
Passing Through the Rings of Saturn
LONNY SCHIFF
A spacecraft passes through the outer layers of the planet.

The Voyager probes achieved their objectives at Jupiter and Saturn and eventually explored all the giant outer planets, specifically their unique systems of rings and magnetic fields, and forty-eight of their moons. They returned information that revolutionized the science of planetary astronomy, helping to resolve some key questions while raising intriguing new ones about the origin and evolution of the solar system. Their discoveries include rings around Jupiter, volcanoes on its moon Io, shepherding satellites in Saturn's rings, new moons around Uranus and Neptune, and geysers on Neptune's moon Triton. The last imaging sequence was the portrait by Voyager 1 of the solar system, depicting Earth and six other planets as sparks in a dark sky lit by a single bright star, the Sun.

It was nearly two decades after Voyager before any spacecraft ventured to the outer solar system again. In October 1989, NASA's Galileo spacecraft began a gravity-assisted journey to Jupiter, where it sent a probe into the atmosphere and observed the planet and its satellites for two years after reaching its destination in December 1995. Jupiter had been of great interest to scientists because it appeared to contain material in its original state left over from the formation of the solar system, and the mission was designed to investigate the chemical composition and physical nature of Jupiter's atmosphere and satellites. But the mission was star-crossed, for soon after

Galileo
HENK PANDER
Prior to launch, the Galileo spacecraft undergoes the Solar Vacuum Test at the Jet Propulsion Laboratory.

OPPOSITE:
Jupiter
LILIKA PAPANICOLAOU
The surface of Jupiter is illuminated in an orange glow.

deployment from the space shuttle, Galileo's umbrellalike, high-gain antenna would not fully open. Through the end of 1995 the spacecraft's performance and condition were excellent, but science and engineering data had to be transmitted via a slower and less effective low-gain antenna.

On December 7, 1995, Galileo dispatched a probe into Jupiter's atmosphere and its instruments relayed data back to the orbiter on chemical composition, the nature of the cloud particles and structure of the cloud layers, the atmosphere's radiative heat balance, pressure and dynamics, and the ionosphere. The probe lasted for about forty-five minutes before the atmosphere and the pressure of the planet destroyed it. During that time Galileo stored the data returned, but with the high-gain antenna inoperative, scientists and technicians spent months retrieving information for analysis.

The results have already begun to reinterpret human understanding of Jupiter and its moons. On August 13, 1996, data from the probe revealed that Jupiter's moon Europa may harbor "warm ice" or even liquid water—key ele-

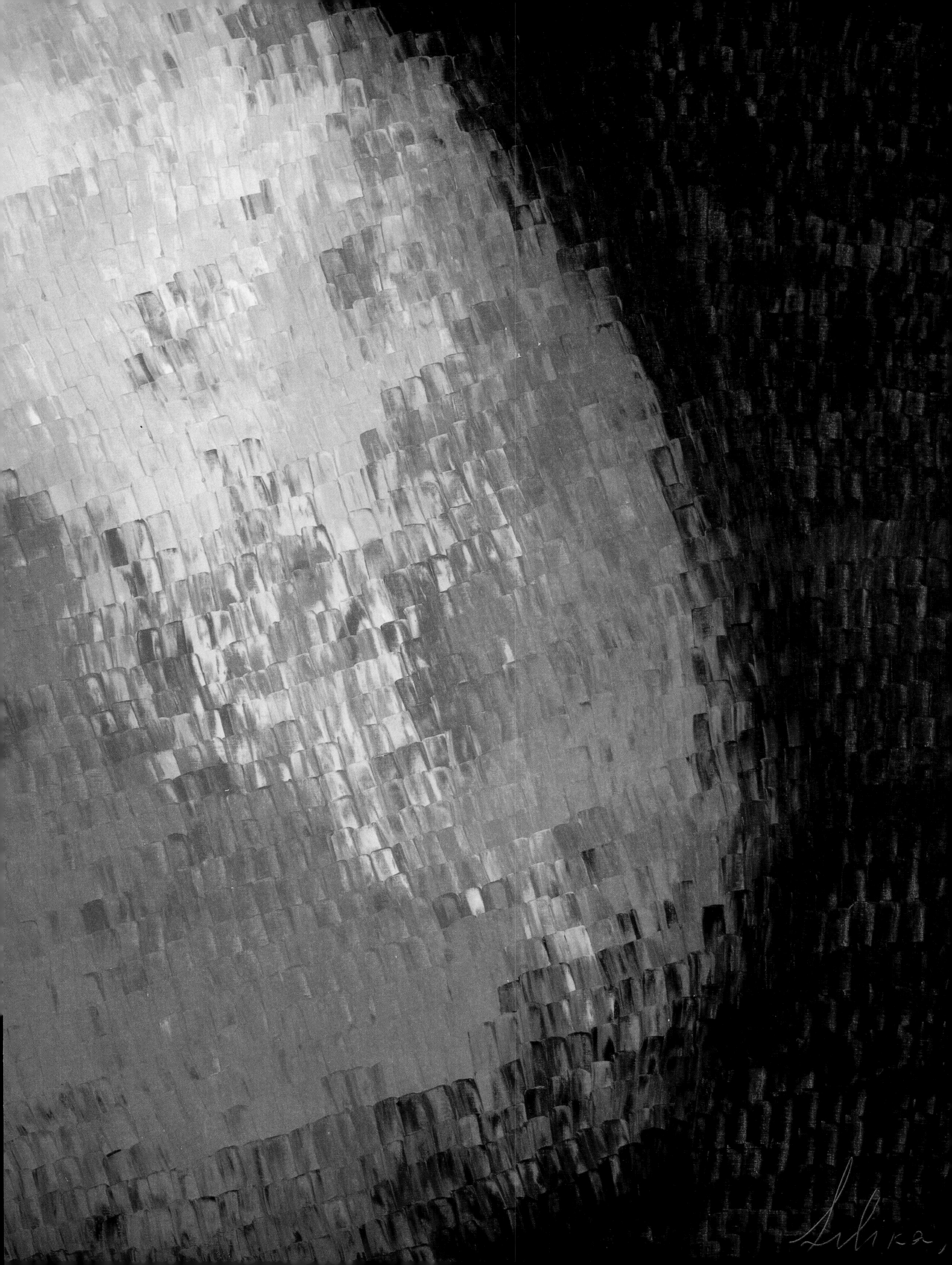

Ammonite: Saturn and Beyond
LAMAR DODD
These three panels link the curves of Saturn with the carved shapes of ammonite fossil shells, which here, possibly represent Earth.

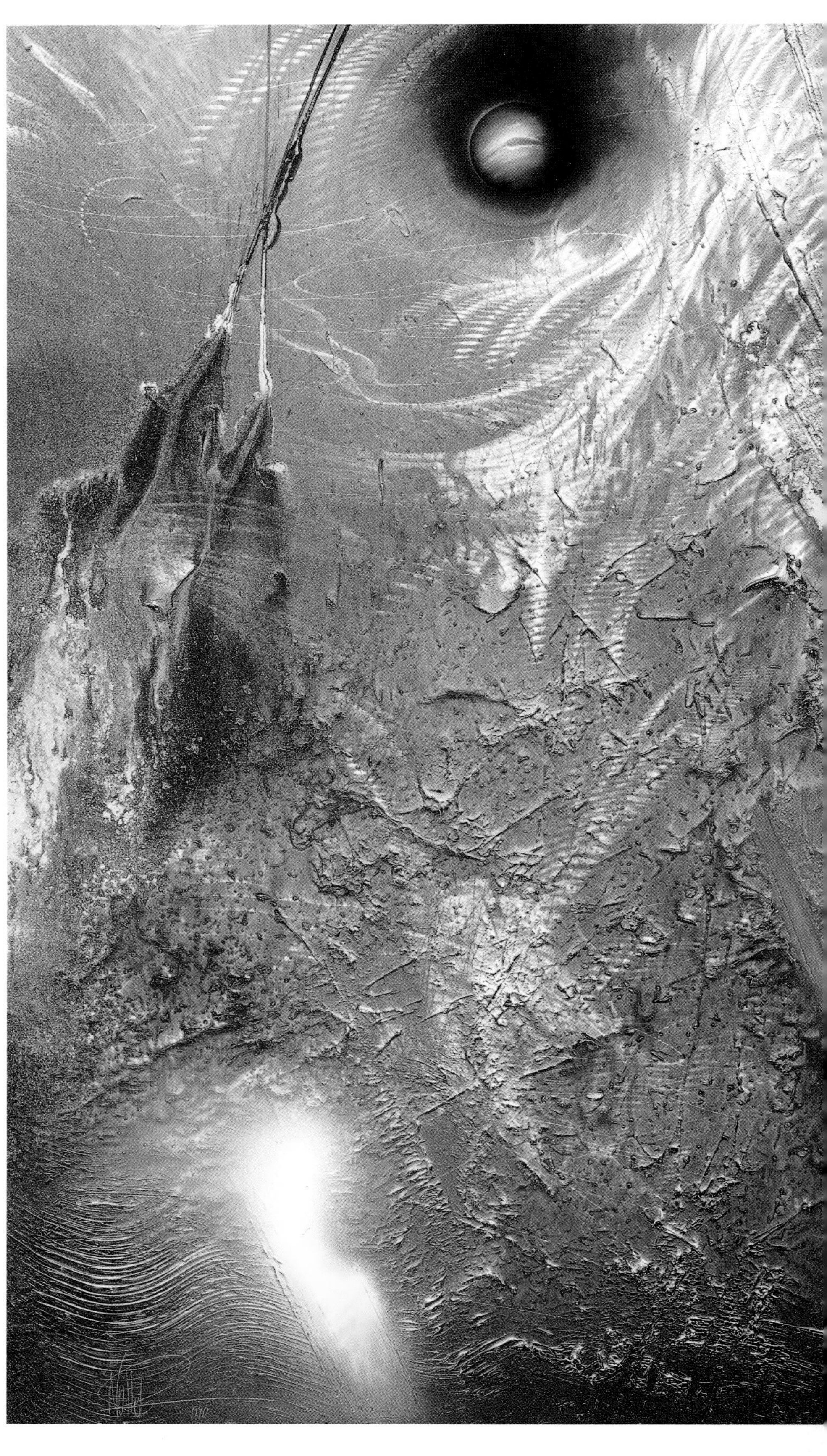

Neptune and the Fire of Knowledge
ANDREAS NOTTEBOHM
In 1989, Voyager 2 journeyed within 3,000 miles of Neptune's north pole. It then swept past the moon Triton and discovered six new moons.

Unexplainable Textures Emerging on Saturn, the 13th of October 2044
ANDREAS NOTTEBOHM

ments in life-sustaining environments. Galileo's photos of Europa, taken during a fly-by some 96,000 miles away, reveal what look like ice floes similar to those seen in the Earth's polar regions. The pictures indicate what appear to be giant cracks in Europa's ice where warm-water "environmental niches" may exist. In early 1997, Galileo also discovered icebergs, lending credence to the possibility of hidden, subsurface oceans. Many scientists and science-fiction writers have speculated that Europa, in addition to Mars and Saturn's moon Titan, is one of the three planetary bodies in this solar system that may possess or has possessed an environment for primitive life. These findings generated new questions in this regard, but NASA scientists stressed that the data was not conclusive and advocated sending a probe to Europa.

Finally, on October 15, 1997, the international Cassini space probe mission left Earth bound for Saturn atop an Air Force Titan IV-B/Centaur rocket from Cape Canaveral, Florida. With the European Space Agency's Huygens

Pioneer 11
WILSON HURLEY
The Pioneer 11 space probe on its voyage toward Saturn.

probe and a high-gain antenna provided by Italy's space agency, Cassini will arrive at Saturn on July 1, 2004, when it will begin an extensive mapping and data-collection effort on the ringed planet.

INVESTIGATING THE UNIVERSE

While inner- and outer-planetary exploration missions were fundamentally reshaping scientific knowledge, space scientists were also profoundly affecting humanity's understanding of the universe beyond. The traditional scientific field of astronomy underwent a tremendous burst of activity with the advent of the space age because of the ability to study stars through new types of telescopes. In addition to greatly enhanced capabilities for observation in the visible light spectrum, NASA and other institutions support the development

Burned Retina
DOUG AND MIKE STARN
An interpretation of NASA's SOHO project to study the Sun.

of a wide range of X-ray, gamma-ray, ultraviolet, infrared, microwave, cosmic ray, radar, and radio astronomical projects. These efforts collectively informed the most systematic efforts yet undertaken to explain the origins and development of the universe.

One of the most important and publicly exciting astronomical activities was the Hubble Space Telescope. Although it was impaired when first placed in Earth orbit in 1990, subsequent corrections to its optics via a shuttle servicing mission in December 1993 have allowed it to return exceptional scientific data about the origins and development of the universe. One of the Hubble's important discoveries took place in early 1997, when NASA scientists announced the discovery of three black holes in three normal galaxies, suggesting that nearly all galaxies may harbor supermassive black holes that once powered quasars (extremely luminous nuclei of galaxies) but are now

Space Telescope Assembly
John Solie
Technicians on various platform levels work on the Hubble Space Telescope.

Deployment
BRYN BARNARD
The Hubble Space Telescope is deployed from the space shuttle's cargo bay using the Remote Manipulator System.

JOHN SOLIE
'94

HUBBLE SERVICING MISSION

Flight day ten was the most relaxed day of the mission. We had been executing a water-dump procedure. (Aboard the orbiter, hydrogen and oxygen are combined to produce electricity and water, more water than the crew can use.) The excess was dumped overboard through a nozzle on the port side just behind the side hatch and the resulting stream of water had broken up into tiny droplets that instantly froze. As the sun rose on our starboard side, the view out of the side-hatch window was one of a blizzard of sun-reflecting ice crystals.

During the previous nine days, we had checked out the mechanical arm and four spacesuits, rendezvoused with the Hubble Space Telescope (HST), performed five spacewalks, and then released the telescope into orbit on its own, slightly above and behind us. It had now been outfitted with new solar arrays to convert the sun's light into electricity, without the jitter characteristic of the original arrays. It also featured a new Wide Field and Planetary Camera, with optics to correct spherical aberration in the primary mirror. A refrigerator-sized silver box had been installed to focus sunlight onto other scientific instruments, and all remaining scheduled upgrades and repairs had been completed. None of our many contingency plans and "work arounds" were executed because every task was accomplished as planned. Truly a miracle, in my opinion.

While ground-control teams were preparing Hubble for its first clear look at the Universe, we were preparing to bring the Space Shuttle Endeavour home. Over 150 tools used during the spacewalks were returned to their proper locations and the spacesuits that had served us well were cleaned and stowed in the airlock for entry. Endeavour's flight controls also had to be checked out before entry day.

"Hey guys! Look at this!" I called out. Through the water-dump blizzard, many miles away, the rising sun was reflecting off the bottom of the HST, which caused it to shine as the brightest of stars. With Christmas music playing on a personal stereo in the background, we all marveled at the beauty of our "Christmas Star," as well as all that had been accomplished in the preceding days. With the frenzy of last-minute details before launch, and the intensity of this very important mission, I had forgotten that it was Christmastime—a season of miracles!

KATHY THORNTON

Servicing the
Hubble Space Telescope,
JOHN SOLIE
As part of the successful repair mission,
astronaut Kathy Thornton
triumphantly releases the solar panel
of the Hubble Space Telescope.

quiescent. This conclusion was based on a census of twenty-seven nearby galaxies carried out by NASA's Hubble Space Telescope and ground-based telescopes in Hawaii, which were used to conduct a spectroscopic and photometric survey of galaxies to find black holes that have consumed the mass of millions of sunlike stars. The key results were: (1) supermassive black holes appear so common that nearly every large galaxy has one; (2) a black hole's mass seems proportional to the mass of the host galaxy, so that, for example, a galaxy twice as massive as another would have a black hole that is also twice as massive; and (3) the number and masses of the black holes found should be consistent with what would have been required to power the quasars.

Less well-known than the Hubble but no less significant, the Cosmic Background Explorer (COBE) was launched on November 18, 1989, from Vandenberg Air Force Base in California. COBE provided the first capability of undertaking an all-sky survey and to measure the cosmic microwave background anisotropy and the interplanetary dust cloud at different solar elongation angles. In essence, these instruments could help to determine the origins and development of the universe at the time of the Big Bang. A major finding was published in April 1992 when COBE scientists announced at the American Physical Society's annual meeting that they had detected long-sought variations within the glow of the Big Bang. This detection represents a major milestone in a twenty-five-year search, and it supports theories explaining how the initial expansion of the universe began fifteen billion years ago.

NASA'S NEXT GENERATION OF SPACE MISSIONS

Because of budgetary and other constraints in the early 1990s, NASA leaders moved toward the construction of a greater number and variety of smaller, inexpensive satellites instead of just a few large, expensive spacecraft. Upon his arrival in 1992, NASA Administrator Daniel S. Goldin urged a philosophy of "smaller, cheaper, faster" for the agency's space missions. What emerged was a relatively dynamic program aimed at discovering the answers to critical scientific and technological questions under a philosophy that capitalizes on the age-old maxim of not placing all one's eggs in a single basket. The Mars Pathfinder mission was a prime example of this philosophy and represents the type of economical program NASA has sought since the end of the Cold War as a measure of its efforts for the twenty-first century.

Instead of developing resource-draining projects defined by narrow sets of parameters, question-driven space exploration efforts of the future might enable genuine understanding about the universe and humanity's place in it. It may also lead to the widespread use of space as possibly the eventual destiny of humankind. In this context, the following questions of discovery are helping to guide planning space exploration missions for the twenty-first century:

How did the universe, galaxies, stars, and planets form and evolve?

How can exploration of the universe and solar system revolutionize the understanding of physics, chemistry, and biology?

Does life in any form, however simple or complex, carbon-based or other, exist elsewhere than on planet Earth?

Are there Earth-like planets beyond our solar system?

How can we utilize knowledge of the Sun, Earth, and other planetary bodies to develop predictive environmental, climate, natural disaster, and

natural resource models, thus helping ensure sustainable development and improvement of the quality of life on Earth?

What is the fundamental role of gravity and cosmic radiation in vital biological, physical, and chemical systems in space, on other planetary bodies, and on Earth, and how can this fundamental knowledge be applied to the establishment of a permanent human presence in space to improve life on Earth?

How can we enable revolutionary technological advances to provide air and space travel for anyone, anytime, anywhere more safely, more affordably, and with less environmental impact, yet still improve business opportunities and global security?

What cutting-edge technologies, processes, and techniques and engineering capabilities must we develop to enable our research agenda to evolve in the most productive, economical, and timely manner?

2092 A.D.
TINA YORK
A futuristic satellite hovers over an abstract road into space, a possible metaphor for a larger spacecraft.

PAGES: 186–87:
View of Earth
DENNIS DAVIDSON
The Moon floats serenely above the Earth's horizon.

How can we most effectively transfer the knowledge gained from research and discovery to commercial ventures in the air, in space, and on Earth?

Perhaps the best question-driven scientific effort underway is the NASA Origins Program: "The Astronomical Search for Our Origins." Guided by a set of fundamental questions, NASA plans the launch of a series of missions beginning in the next century that are designed to help answer age-old astronomy riddles. Origins missions will include ground-based telescopes and four space-based observatories.

The Next Generation Space Telescope (NGST) will be a follow-on observatory to the Hubble Space Telescope, consisting of a large single telescope folded to fit inside its launch vehicle and cooled to low temperatures in deep space to enhance its sensitivity to faint, distant objects. NGST will be up to twenty-five feet across, much larger than even the space shuttle could carry in one piece, and it will use new technologies for automatic deployment after launch. Its mirror will be very thin and adjusted to the proper shape after it is cooled down to operating temperature. Placed in an orbit far from Earth, away from the heat radiated by our home planet, its operating temperature will be low enough so that even its own thermal energy does not swamp faint astronomical signals.

The Terrestrial Planet Finder (TPF) will consist of a collection of four small telescopes, each about eighty inches across, precisely located on a 240-foot-long truss and with the beam of their lights set to a common focus. With the telescopes functioning together, resulting images would be as sharp as that of a single telescope the size of a football field. The infrared light collected from all four telescopes will be carefully combined so that starlight is rejected but planetary light is collected and analyzed, thus enabling a search beyond this solar system for planets like Earth—a daunting task, for the challenge of finding such a planet orbiting even the closest stars can be compared to finding a tiny firefly next to a blazing searchlight thousands of miles away.

The Space Infrared Telescope Facility (SIRTF), to be launched in 2001, will overlap the operations of Hubble and complement its observations. SIRTF will answer questions about the dust clouds surrounding stars, which may mark the sites of developing planets, and its sophisticated cooled optics will increase our knowledge of planet, star, and galaxy formation.

Space Interferometry Mission (SIM), to be launched early in the next decade, will allow scientists to "take the measure of the universe" by providing absolute positions of stars with the precision of a few millionths of an arcsecond, conceivably allowing someone on Earth to see a human standing on the Moon while switching a flashlight from hand to hand. It will map the "wobbles" of nearby stars as their paths weave across the sky, providing indirect evidence that these stars have planets orbiting around them which exert gravitational pull. From these observations, astronomers will be able to infer the presence of planets as small as the Earth itself. These measurements will establish the first rungs of the cosmic distance "ladder" by which astronomers judge distances to all objects in the universe.

Collectively, these four new astronomical observatories will offer enormously exciting possibilities for future generations to understand more fully the universe in which they live.

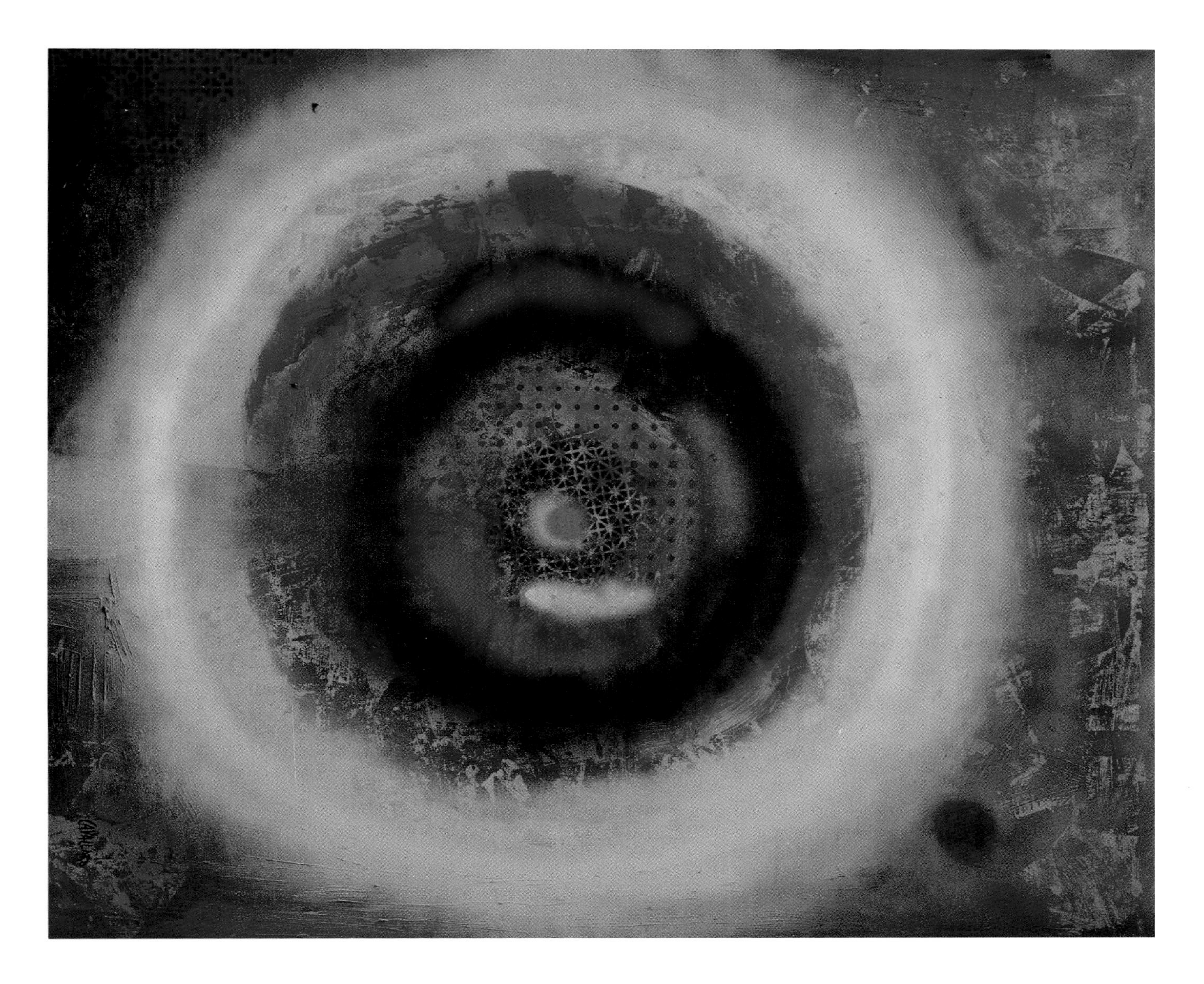

T + 30 Seconds
Vincent Cavallero

Pages 191–92:
Mission to Mars
Ren Wicks
A visionary depiction of astronauts exploring the Martian terrain.

A NEW GENERATION ON THE MOON AND MARS

Lunar and planetary exploration, by both robotic probes and humans, also hold tremendous promise for the next century. The lure of the Moon and Mars remains seductive for both those involved in the field and the public at large. Excitement abounds when missions to these heavenly bodies are undertaken. For instance, NASA's "New Millennium" program is designed to develop new microtechnology and advanced automation techniques, precipitating a revolution in the design and operation of scientific spacecraft. A fleet of these small, high-tech probes will return a continuous flow of information to Earth about the solar system and possibly even the existence of planets around nearby stars.

The centerpiece of this effort is the already-underway Mars Surveyor effort, a ten-year, multispacecraft program in which two small science probes will be sent to explore Mars in 1999 to demonstrate the innovative new technologies of NASA's New Millennium program. Microprobes will hitchhike to Mars aboard NASA's 1998 Mars Surveyor lander, to be launched in January 1999, and will complement the climate-related scientific focus of the lander by demonstrating an advanced, rugged microlaser system for detecting sub-

NASA
NASA
NASA
NASA

Mars from Deimos
LUDEK PESEK
Mars is visible from the surface of the moon Deimos.

surface water. Such data on polar subsurface water, in the form of ice, should help limit scientific projections for the global abundance of water on Mars. Future missions to the planet could use similar penetrators to search for subsurface ice and minerals that could contribute to the search for evidence of life on Mars.

These spacecraft will spend eleven months en route to the Red Planet. Just prior to their entry into the Martian atmosphere, the microprobes, mounted on the spacecraft's cruise ring, will separate and plummet to the surface using a single-stage entry aeroshell system. The probes will plunge into

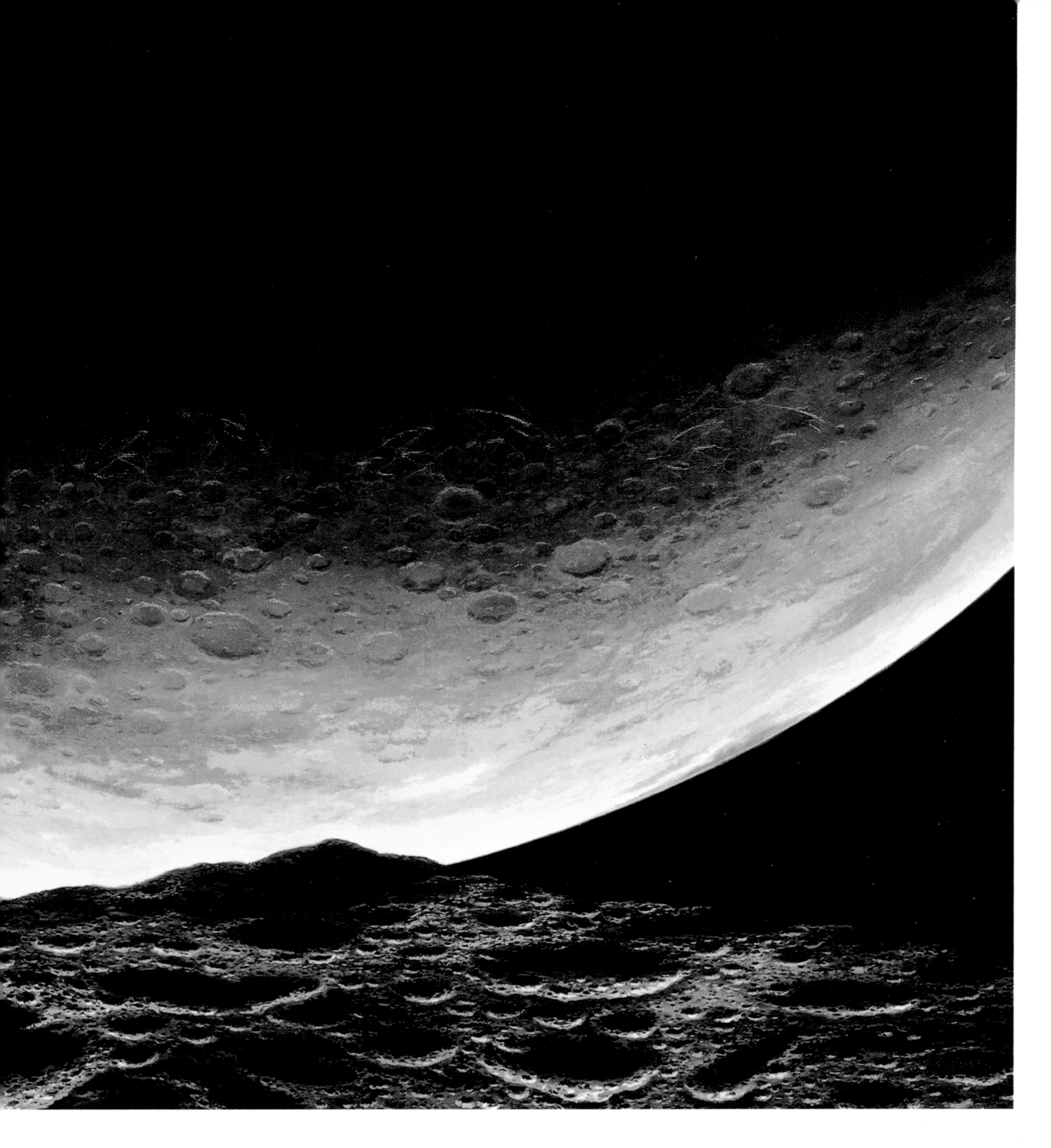

the surface of Mars at an extremely high velocity of about 446 miles per hour (200 meters per second) to ensure maximum penetration of the Martian terrain. They should impact the surface within 120 miles of the main Mars lander, which is targeted for the planet's icy south polar region. The microprobes will weigh less than four and a half pounds each and be designed to withstand both very low temperatures and high deceleration. In-situ instrument technologies for making direct measurements of the Martian surface will include a water and soil sample experiment, a meteorological pressure sensor, and temperature sensors for measuring the thermal properties of the Martian soil.

Earth and Moon
GREG MORT
A symbolic union of the Earth and the Moon.

PAGE 196:
Unlocking Venus's Secrets
LONNY SCHIFF

PAGE: 197:
Viking 2 Passing Over Mars
LONNY SCHIFF
A symbolic interpretation of the the Viking Lander passing over Mars's surface on November 2, 1982.

© Lonny Schiff '89

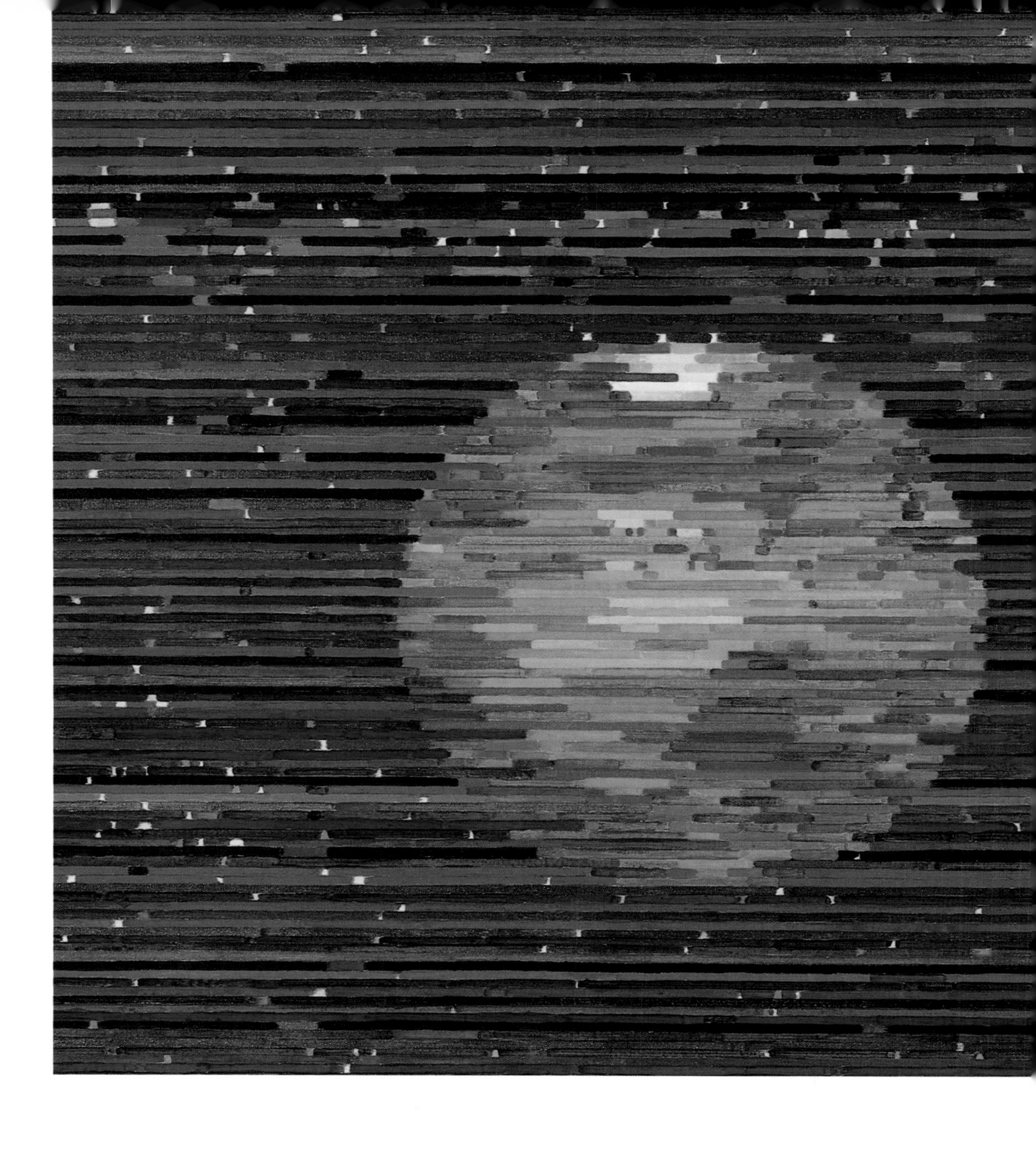

TRAVELING THE SOLAR SYSTEM

The core mission of any future space exploration will be human departure from Earth orbit and the journey to the Moon or Mars for extended and perhaps permanent stays. Active efforts to develop both the technology and the scientific knowledge necessary to carry out this mission are well underway in the form of a new generation of launch vehicles currently being built. The X-33, X-34, and other hypersonic research projects will help to realize routine and affordable access to space during the first decades of the twenty-first century.

This idea of routine and low-cost space access is the key to expanding research and exploration, as well as unleashing the great commercial potential of space. Through integration of aeronautical principles with commercial launch vehicles, a ten-fold reduction in the cost of placing payloads in low-Earth orbit are now being projected by the year 2010, and an additional ten-fold cost reduction during the decade beyond is a long-term goal. Through the X-33 and other future launch systems, NASA envisions the space frontier

Mars Expedition
Michael Pranzatelli
Planet Mars and its two small moons, Phobos and Deimos, await further discovery.

as a busy crossroads of American-led international science, research, commerce, and exploration.

In this context, the potential for the future seems almost limitless. Putting a human on Mars, which once seemed like science fiction, is rapidly becoming a possibility. NASA Administrator Goldin remarked, "It's not a question of *if* it's going to happen. It's a question of *when.*" Goldin has suggested that NASA conduct up to ten missions to Mars in the next ten years. The objectives of these missions will range from determining the physical resources of Mars to converting Mars's carbon dioxide atmosphere to oxygen.

A defined set of missions has already been slated, with the overarching goal being to answer, once and for all, the fundamental question of whether life ever did exist on Mars. In contrast to the Viking missions of the 1970s, NASA's current Mars Exploration Program focuses on past life and, therefore, primarily on the earliest epochs of Martian history when hydrological processes played a large part in shaping its surface. The ongoing 1996 and 1998 orbiter and lander missions are directed toward: (1) a systematic map-

PAGES 200–201:
Chaos and Order
James Cunningham
Seen here is Uranus with the scarp of its moon Miranda in the foreground.

ping of the Martian surface to establish a basic characterization of the planet, (2) a study of the atmosphere and of polar volatiles, and (3) the development of technology to be used by future landers. Beginning in 2001, a series of missions will complete systematic global mapping from orbit (2001 Mars Surveyor Orbiter); detailed surface exploration of the most promising exobiology target area (2001 and 2003 Mars Rovers); and the collection and return of samples from a targeted exobiology Martian site (2005 Mars Sample Return).

The human exploration of Mars is a long-term goal that appeals to virtually everyone, and one of the most important efforts to achieve it emanated from an August 1992 workshop of space-flight experts, convened under NASA's direction but involving many other organizations. It represents one chapter in the ongoing process of melding new and existing technologies, practical operations, fiscal reality, and common sense into a feasible and viable human mission to Mars. Far from the final word on this planning process, the workshop emphasized three objectives for the analysis of a Mars exploration program and the first piloted missions in that program, and it involved: (1) human missions to Mars verifying how humans can ultimately inhabit Mars, (2) applied science research to determine how Mars's resources can augment life-sustaining systems, and (3) basic science research on the solar system's origin and history.

The eventual human missions to Mars that will be required to accomplish future exploration and research activities will also call for safe transportation, maintenance on the surface of Mars, and the return to Earth of a healthy crew. Investigations in these areas are ongoing, with a human mission to Mars possible in the first third of the twenty-first century.

No matter how Martian exploration originates, it is not a short-term venture; it will require not only technological and scientific know-how but also the political will of both the American people and their elected leaders. It will become an enduring effort, one not unlike the settlement of the Americas in the period between 1500 and the present. The first expeditions will be halting, discovery-oriented missions, but later trips will eventually entail the establishment of colonies occupied by people possessing the skills necessary for living on Mars. Toward the end of the next century, humans may build small factories, using supplies from Earth, and rely upon Mars's natural resources to manufacture other necessities such as additional building materials.

CONCLUSION

The exploration of space in the next century promises enormous returns. The world is embarking on an exciting new era in space exploration and utilization. New technologies will enable faster, lower-cost mission design and development. Miniaturization and integration will allow smaller, lower-mass spacecraft, resulting in savings in total mission costs. Ground operations, including launch services, ground communications, mission operations, and scientific-data processing, are also receiving extended efforts to reduce costs and yield higher returns. The result is that in the twenty-first century, space flight may become a much more common phenomenon. Like the rise of airplane travel, realizable space travel holds promise not for just the distant

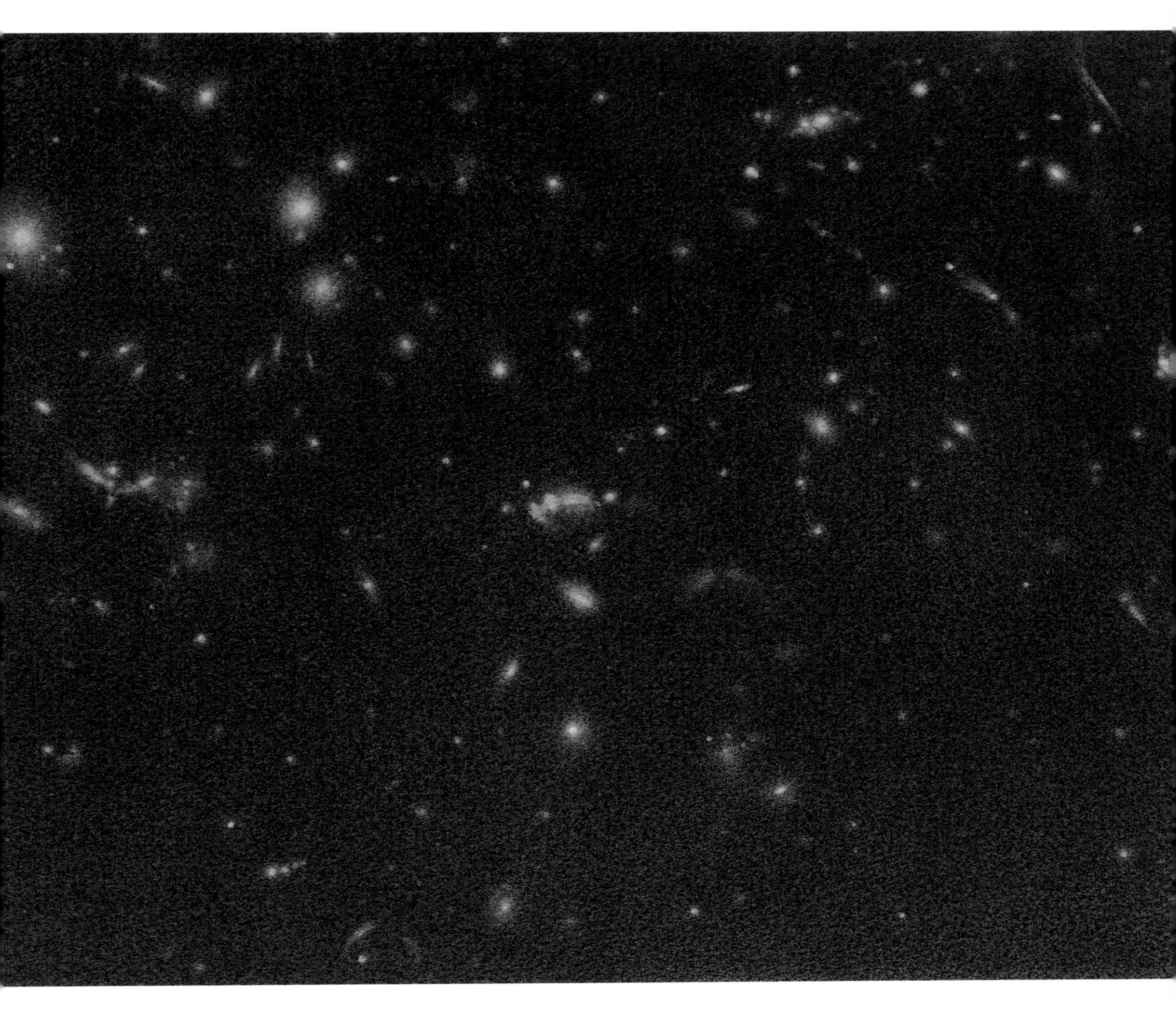

future but for the next fifty years. America will become a spacefaring nation just as surely as it became an airfaring nation. One hundred years from now, historians will probably look back at the end of this century with the same curiosity and perception of quaintness with which current researchers consider the efforts of the Wright brothers just under a hundred years ago. What a difference a century makes!

View of the Universe Five to Twelve Billion Light Years Away
Vija Celmins

Shuttle Flowers
James Dean
Fields of wildflowers in the foreground of
Launch Complex 39A at the Kennedy Space Center.

After Touchdown
Robert Schulman
Post-flight desolation the day after a shuttle landing at
the Dryden Research Facility in 1981.

OPPOSITE:
Emergence
Dan Namingha
An astronaut and Native American spirits float together in space.

THE NASA ART PROGRAM

Imagine sitting on the deck of a recovery ship awaiting the splashdown of a Mercury astronaut or in the cockpit of the space shuttle, hours after its return from the heavens. Artists have watched Neil Armstrong during suit-up, discussed fluid dynamics with NASA engineers, journeyed to Russia to witness Apollo-Soyuz negotiations, and observed scientists tinker with delicate instruments for a space station. Some have even witnessed one of NASA's bleakest moments, the Challenger tragedy.

Beginning in 1963, under the leadership of NASA Administrator James Webb, NASA embarked on an ambitious plan to document major events through the eyes of artists. James Dean, an artist himself, was chosen to run the NASA Art Program with the help of Lester Cooke, Curator of the National Gallery. The 1960's proved to be essentially a magical era for the space agency. As NASA scientists, engineers, and astronauts increased the chances of achieving their lunar goal, American popular culture became enthralled with space.

This excitement spilled into art circles as well. As a result, NASA was able to invite a number of reputable artists, including Robert Rauschenberg, Norman Rockwell, and James Wyeth, to render NASA activities artistically. Although the invitation included a very meager honorarium (and still does), it nonetheless provided access to some of the most exciting moments that have defined America's history.

From the late 1970's until the early 1990's, artist Robert Schulman managed and curated the

NASA Art Program. Although many earlier works were donated to the National Air and Space Museum when it opened in 1976, new commissions did not diminish. In fact, the program became a very efficient apparatus. Actual "art teams" were sent to various NASA installations to chronicle a particular NASA event. As a result, NASA soon began building an impressive collection, mostly dedicated to the hallmark of the era: the space shuttle.

Although fewer commissions are offered today, there has been an effort to cover new areas of NASA's scientific efforts, ranging from environmental studies to state-of-the-art planetary probes. With the help of galleries and museums such as Leo Castelli, the McKee Gallery and the Studio Museum of Harlem, NASA has been able to commission artists like Vija Celmins and the Starns. Venues for exhibiting NASA artwork have included the National Gallery of Art, the Smithsonian, and many other museums and galleries around the country.

NASA is as much a show-stopping display of scientific achievement as it is a celebration of human effort. The NASA Art Program is certainly an extension of the latter—and more personal—element. Although some of the works are more formidable than others, there remains in most the indelible mark of human inspiration. There are parallels between artist, scientist, and astronaut. They are all humans discovering new ways to interpret the unknown.

In measuring the value of the NASA Art Program, one should balance the modest honorarium fee awarded to artists with the benefit of what is finally created. As a whole, the works comprise a significant collection representing a variety of styles and subjects. Not only do these works continue to inspire, they provide a historical legacy for future generations to behold.

Bertram Ulrich
Curator, NASA Art Program

Study for Blockhouse (see pages 6-7)
Fred Freeman

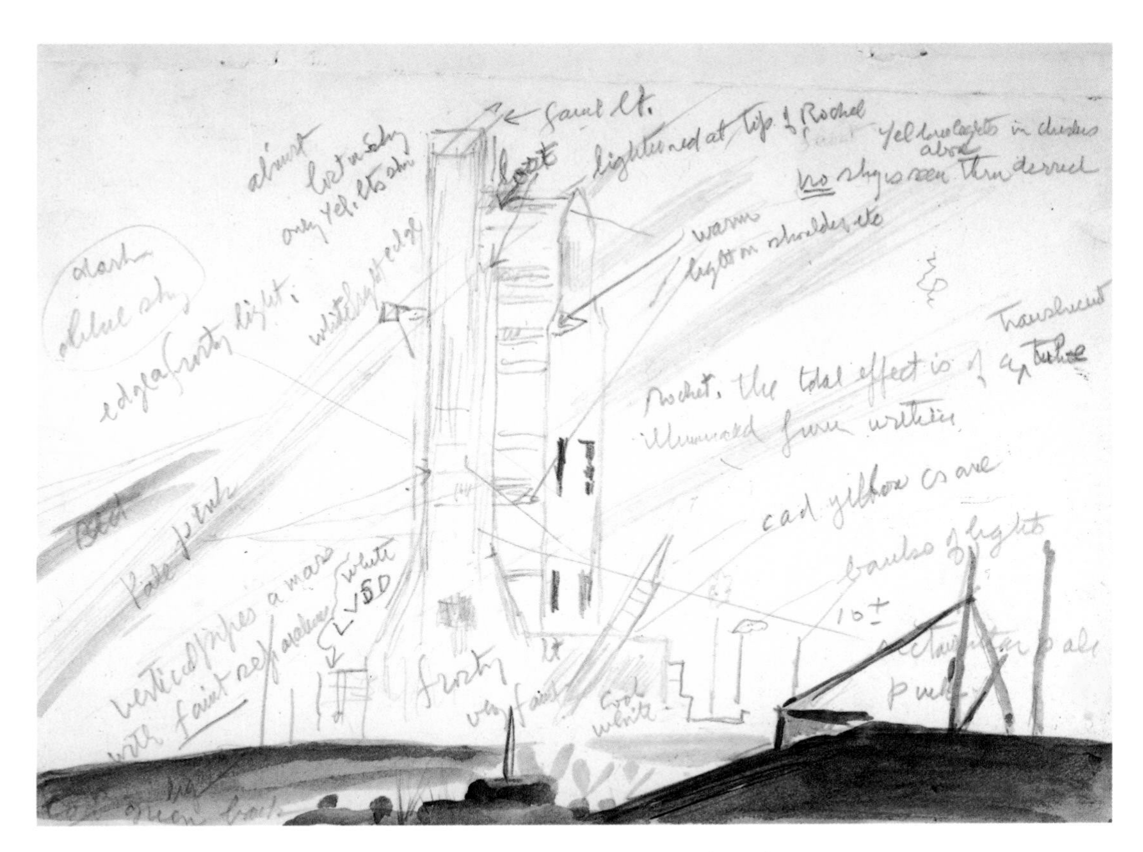

Preliminary Sketch of Launch Pad
PETER HURD

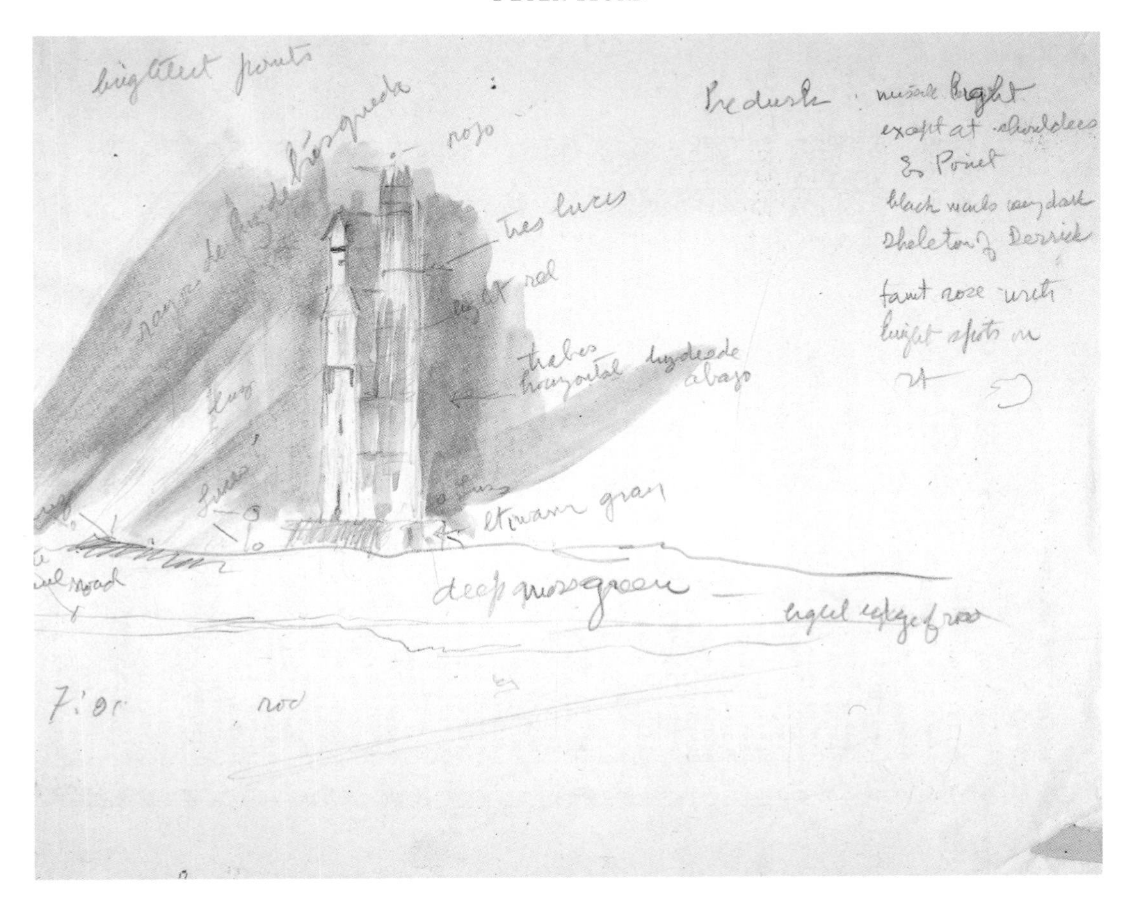

Watercolor Sketch of Skylab
PETER HURD

MERCURY FLIGHTS
1961–1963

SPACECRAFT	LAUNCH DATE	CREW	FLIGHT TIME			HIGHLIGHTS
			Days	Hours	Min.	
Mercury-Redstone 3	5 / 5 / 61	Alan B. Shepard	0	0	15	First U.S. flight; suborbital
Mercury-Redstone 4	7 / 21 / 61	Virgil I. Grissom	0	0	16	Suborbital; Redstone capsule sank following landing; Grissom recovered
Mercury-Atlas 6	2 / 20 / 62	John H. Glenn	0	4	55	First American to orbit
Mercury-Atlas 7	5 / 24 / 62	M. Scott Carpenter	0	4	56	Landed 400km (880 miles) from target
Mercury-Atlas 8	10 / 3 / 62	Walter M. Schirra	0	9	13	Landed 8km (18 miles) from target
Mercury-Atlas 9	5 / 15 / 63	L. Gordon Cooper	1	10	20	First U.S. flight exceeding 24 hours

GEMINI FLIGHTS
1965–1966

SPACECRAFT	LAUNCH DATE	CREW	FLIGHT TIME			HIGHLIGHTS
			Days	Hours	Min.	
Gemini 3	3 / 23 / 65	Virgil I. Grissom John W. Young	0	4	53	First U.S. 2-person flight; first manual orbital maneuvers
Gemini 4	6 / 3 / 65	James A. McDivitt Edward H. White	4	1	6	21–min. extravehicular activity (White)
Gemini 5	8 / 21 / 65	L. Gordon Cooper Charles Conrad	7	22	55	Longest human flight to date
Gemini 7	12 / 4 / 65	Frank Borman James A. Lovell, Jr.	13	18	35	Longest human flight to date
Gemini 6-A	12 / 15 / 65	Walter M. Schirra Thomas P. Stafford	1	1	51	Rendezvous within 30 cm. of Gemini 7
Gemini 8	3 / 16 / 66	Neil A. Armstrong David R. Scott	0	10	41	First docking of 2 orbiting spacecraft (Gemini 8 with an Agena target vehicle)
Gemini 9-A	6 / 3 / 66	Thomas P. Stafford Eugene A. Cernan	3	0	21	Extravehicular activity; rendezvous

GEMINI FLIGHTS
1965–1966

SPACECRAFT	LAUNCH DATE	CREW	FLIGHT TIME			HIGHLIGHTS
			Days	Hours	Min.	
Gemini 10	7 / 18 / 66	John W. Young Michael Collins	2	22	47	First dual rendezvous (with Agena 8 and 10)
Gemini 11	9 / 12 / 66	Charles Conrad, Jr. Richard F. Gordon, Jr.	2	23	17	First initial-orbit docking; first tethered flight; highest Earth-orbit altitude (1,372km/3,018 miles)
Gemini 12	11 / 11 / 66	James A. Lovell, Jr. Edwin E. Aldrin, Jr.	3	22	35	Longest EVA to date (Aldrin, 5:37:00)

APOLLO FLIGHTS
1967–1972

SPACECRAFT	LAUNCH DATE	CREW	FLIGHT TIME			HIGHLIGHTS
			Days	Hours	Min.	
Apollo 7	11 / 11 / 68	Walter M. Schirra, Jr. Walter Cunningham Donn Eisele	10	20	09	First U.S. 3-person mission
Apollo 8	12 / 21 / 68	Frank Borman James A. Lovell, Jr. William A. Anders	6	3	01	First human orbit(s) of Moon; first human departure of Earth's sphere of influence; highest speed attained in human flight to date
Apollo 9	3 / 3 / 69	James A. McDivitt David R. Scott Russell L. Schweickart	10	1	01	Simulation by LM of lunar landing, takeoff, and rejoining to command module while in Earth orbit
Apollo 10	5 / 18 / 69	Thomas P. Stafford John W. Young Eugene A. Cernan	8	0	30	Successful demo of complete system (with LM) to 14,300m from lunar surface.
Apollo 11	7 / 16 / 69	Neil A. Armstrong Michael Collins Edwin E. Aldrin, Jr.	8	3	09	First manned landing on Moon
Apollo 12	11 / 14 / 69	Charles Conrad, Jr. Richard F. Gordon, Jr. Alan L. Bean	10	4	36	Second manned lunar landing. Exploration of lunar surface. Parts retrieval of Surveyor 3*

*which landed in Ocean of Storms on 4/19/67.

APOLLO FLIGHTS
1967–1972

SPACECRAFT	LAUNCH DATE	CREW	FLIGHT TIME			HIGHLIGHTS
			Days	Hours	Min.	
Apollo 13	4 / 11 / 70	James A. Lovell, Jr. John L. Swigert, Jr. Fred W. Haise, Jr.	5	22	55	Mission aborted after explosion in service module. Ship circled Moon with crew using LM as "lifeboat" until just before reentry.
Apollo 14	1 / 31 / 71	Alan B. Shepard Stuart A. Roosa Edgar D. Mitchell	9	0	20	Third manned lunar landing. Mission demonstrated pinpoint landing capability and continued human exploration.
Apollo 15	7 / 26 / 71	David R. Scott Alfred M. Worden James B. Irwin	12	7	12	Fourth manned lunar landing and first Apollo "J–series mission," which carried Lunar Roving Vehicle. Worden's in-flight EVA of 38 min., 12 sec. was performed during return trip.
Apollo 16	4 / 16 / 72	John W. Young Charles M. Duke, Jr.	11	1	51	Fifth manned lunar landing and second with Lunar Roving Vehicle (LRV)
Apollo 17	12 / 07 / 72	Eugene A. Cernan Harrison H. Schmitt Ronald E. Evans	12	13	52	Sixth and final Apollo manned lunar landing, also with LRV

SIGNIFICANT SPACE SHUTTLE FLIGHTS 1981–1998

SPACECRAFT	LAUNCH DATE	CREW	FLIGHT TIME			HIGHLIGHTS
			Days	Hours	Min.	
Space Shuttle Columbia STS–1	4 / 12 / 81	John W. Young Robert L. Crippen	2	6	21	First flight of Space Shuttle Columbia tested space craft in orbit. First landing of airplanelike craft from orbit for reuse
Space Shuttle Challenger STS–7	6 / 8 / 83	Robert L. Crippen Frederick H. Hauck John M. Fabian Sally K. Ride Norman T. Thagard	6	2	24	Seventh flight of Space Shuttle, launched 2 commercial satellites (Anik C-2 and Palapa B-1), also and retrieved SPAS 01; first flight with 5 crew members including first woman U.S. astronaut.
Space Shuttle Challenger STS–8	8 / 30 / 83	Richard H. Truly Daniel C. Brandenstein Dale A. Gardner Guion S. Bluford, Jr. William E. Thornton	6	1	9	Eighth flight of Space Shuttle, launched one commercial satellite (Insat 1-B), first flight of U.S. African-American astronaut.
Space Shuttle Challenger STS-51L	1 / 28 / 86	Francis R. Scobee Michael J. Smith Judith A. Resnik Ronald E. McNair Ellison S. Onizuka Gregory B. Jarvis Christa McAuliffe	0	0	73	Twenty-fourth STS flight. Because of failure of the solid rocket boosters, due to O-ring blowback, this mission was lost along with the crew 73 seconds into the launch.
Space Shuttle STS–26	9 / 29 / 88	Frederick H. Hauck Richard O. Covey John M. Lounge David C. Hilmers George D. Nelson	4	1	0	Twenty-sixth STS flight. This mission marked the return to flight of the Space Shuttle after a hiatus of more than two years following the loss of Challenger. Launched TDRS 3 communications satellite.
Space Shuttle STS-31	4 / 24 / 90	Loren J. Shriver Charles F. Bolden, Jr. Steven A. Hawley Bruce McCandless II Kathryn D. Sullivan	5	1	16	Thirty-fifth STS flight. Launched Hubble Space Telescope (HST).

SIGNIFICANT SPACE SHUTTLE FLIGHTS 1981–1998

SPACECRAFT	LAUNCH DATE	CREW	FLIGHT TIME			HIGHLIGHTS
			Days	Hours	Min.	
STS-61	12 / 2 / 93	Richard O. Covey Kenneth D. Bowersox Tom Akers Jeffrey A. Hoffman Kathryn C. Thornton	10	19	58	Fifty-ninth STS flight. Restored planned scientific capabilities and reliability of the Hubble Space Telescope.
STS–59	2 / 3 / 95	Claude Nicollier F. Story Musgrave James D. Wetherbee Eileen M. Collins Bernard A. Harris, Jr. Michael Foale Janice E. Voss Vladimir G.Titov (RSA)	8	6	28	Sixty-seventh STS flight. First close encounter in nearly 20 years with Russian spacecraft, close flyby of Space Station Mir.
STS–71	6 / 27 / 95	Robert L. Gibson Charles J. Precourt Ellen S. Baker Gregory Harbaugh Bonnie J. Dunbar Anatoly Y. Solovyev (RSA)(Left on Mir) Nikolai M. Budarin (RSA)(Left on Mir) Vladimir N. Dezhurov (RSA)(Returned from Mir) Gennady M. Strekalov (RSA)(Returned from Mir) Norman Thagard (Returned from Mir)	9	19	22	Sixty-ninth STS flight. Docked with Mir and exchanged crews.
Space Shuttle Atlantis STS–79	9 / 16 / 95	William F. Readdy Terrence W. Wilcutt Jerome Apt Thomas D. Akers Carl E. Walz John E. Blaha Shannon W. Lucid	10	3	19	Seventy-ninth STS flight. Docked with Mir space station. Picked up astronaut Shannon Lucid and dropped off astronaut John Blaha.
Space Shuttle Discovery STS79	2 / 11 / 97	Kenneth D. Bowersox Scott J. Horowitz Mark C. Lee Steven A. Hawley Gregory J. Harbaugh Specialist Steven L. Smith Joseph R. Tanner	11	2	14	Eighty-second STS mission. Second Hubble Space Telescope servicing mission.

SIGNIFICANT SPACE SHUTTLE FLIGHTS 1981–1998

SPACECRAFT	LAUNCH DATE	CREW	FLIGHT TIME			HIGHLIGHTS
			Days	Hours	Min.	
Space Shuttle Columbia STS–94	7 / 1 / 97	James D. Halsell Susan L. Still Janice E. Voss Donald A. Thomas Michael L. Gernhardt Specialist Roger Crouch Greg Linteris	16	12	56	Eighty-fifth STS Mission. 23rd Flight OV-102, Columbia, Microgravity Science Lab-1 Reflight.
Space Shuttle Atlantis STS–86	9 / 25 / 97	James D. Wetherbee Michael J. Bloomfield Vladimir G. Titov (RSA) Scott E. Parazynski Jean-Loup J. M. Chretien (CNES) Wendy B. Lawrence Specialist David A. Wolf	11	14	27	Eighty-seventh STS Mission. 7th Mir docking mission, exchanging crew members.

EARLY ASTRONAUT SELECTIONS

GROUP 1: SELECTED APRIL 9, 1959

M. Scott Carpenter (USN)
L. Gordon Cooper (USAF)
John H. Glenn, Jr. (USMC)
Virgil I. "Gus" Grissom (USAF)
Walter M. Schirra, Jr. (USN)
Alan B. Shepard, Jr. (USN)
Donald K. "Deke" Slayton (USAF)

GROUP 2: SELECTED SEPTEMBER 17, 1962

Neil A. Armstrong (civilian)
Frank Borman (USAF)
Charles "Pete" Conrad (USN)
James A. Lovell, Jr. (USN)
James McDivitt (USAF)
Elliott See (civilian)
Thomas Stafford (USAF)
Edward White II (USAF)
John Young (USN)

GROUP 3: SELECTED OCTOBER 8, 1963

Edwin "Buzz" Aldrin
(USAF, Ph.D. astronautics)
William L. Anders
(USAF, M.S. engineering)
Charles Bassett II (USAF)
Alan Bean (USN)
Eugene Cernan
(USN, M.S. engineering)
Roger B. Chaffee (USN)
Michael Collins (USAF)
R. Walter Cunningham
(USMC, M.S. physics)
Donn Eisele
(USAF, M.S. engineering)
Theodore Freeman
(USAF, M.S. engineering)
Richard Gordon (USN)
Russell Schweickart
(civilian, M.S. astronautics)
Clifton Williams, Jr.
(USMC)

GROUP 4: SELECTED JUNE 28, 1965

Owen Garriott
(civilian, Ph.D. engineering)
Edward Gibson
(civilian, Ph.D. engineering)
Duane Graveline
(civilian, M.D.)
Joseph Kerwin
(USN, M.D.)
F. Curtis Michel
(civilian, Ph.D. physics)
Harrison "Jack" Schmitt
(civilian, Ph.D. geology)

GROUP 5: SELECTED APRIL 4, 1966

Vance Brand (civilian)
John Bull (USN)
Gerald Carr
(USMC, M.S. engineering)
Charles Duke (USAF)
Joseph Engle (USAF)
Ronald Evans
(USN, M.S. engineering)
Edward Givens, Jr. (USAF)
Fred Haise, Jr. (civilian)
James Irwin (USAF, M.S. engineering)
Don Lind (civilian, Ph.D. physics)
Jack Lousma
(USMC, M.S. engineering)
Thomas K. Mattingly II (USN)
Bruce McCandless II
(USN, M.S. engineering)
Edgar Mitchell
(USN, Ph.D. aeronautics and astronautics)
William Pogue
(USAF, M.S. mathematics)
Stuart Roosa (USAF)
John Swigert, Jr.
(civilian, M.S. aerospace science)
Paul Weitz
(USN, M.S. engineering)
Alfred Worden
(USAF, M.S. engineering)

GROUP 6: AUGUST, 4, 1967

Joseph Allen
(civilian, Ph.D. physics)
Philip Chapman
(civilian, Ph.D. instrumentation)
Anthony England
(civilian, M.S. physics)
Karl Henize
(civilian, Ph.D. astronomy)
Donal Holmquest (civilian, M.D.)
William Lenoir
(civilian, Ph.D. engineering)
John Llewellyn
(civilian, Ph.D. chemistry)
F. Story Musgrave
(civilian, Ph.D. chemistry)
Brian O'Leary
(civilian, Ph.D. astronomy)
Robert Parker
(civilian, Ph.D. astronomy)
William Thornton (civilian, M.D.)

GROUP 7: AUGUST 14, 1969

Karol Bobko (USAF)
Robert L. Crippen (USN)
Charles Fullerton
(USAF, M.S. engineering)
Henry Hartsfield (USAF)
Robert Overmyer
(USMC, M.S. astronautics)
Donald Peterson
(USAF, M.S. engineering)
Richard H. Truly
(USN)

NASA FACILITIES

NASA HEADQUARTERS, WASHINGTON, D.C. was established at the time of the agency's activation on October 1, 1958. This site is charged with oversight of all NASA programs and is directed by an administrator who is appointed by the president and confirmed by the Senate.

AMES RESEARCH CENTER, MOFFETT FIELD was named for Joseph S. Ames, longtime National Advisory Committee for Aeronautics (NACA) chairman. Established in San Francisco, California, in 1940 as an aeronautical research facility, it is near the West Coast aeronautical industry. After becoming a part of NASA on October 1, 1958, this center continued to concentrate on aeronautical research but also expanded its efforts to include biomedicine, computer technology, and control systems.

DRYDEN FLIGHT RESEARCH CENTER is located in Mojave, California. Established as the High Speed Flight Research Facility in 1946, it became a part of NASA in 1958 and has been the central location where flight research has taken place ever since.

GODDARD SPACE FLIGHT CENTER in Greenbelt, Maryland, was named for Robert H. Goddard upon its establishment in 1959, when the space research projects of the Naval Research Laboratory were transferred en masse from Navy facilities to NASA. The center has been dedicated to space science and Earth observation ever since.

JET PROPULSION LABORATORY is located in Pasadena, California, and was transferred to NASA from the Army in 1959. It remains a center of space science for the space agency.

JOHNSON SPACE CENTER was originally established by NASA in 1962 as the Manned Spacecraft Center in Houston, Texas, and renamed for Lyndon B. Johnson in 1973. The Space Task Group moved there from Langley Research Center in 1962 and the site is the home of NASA's human space-flight program.

KENNEDY SPACE CENTER was originally an Air Force launch complex near Titusville, Florida. KSC was an early NASA facility. Named for John F. Kennedy in 1963, it has been NASA's spaceport ever since.

LANGLEY RESEARCH CENTER was named for Samuel P. Langley and established in 1917, the earliest of the NACA's aeronautical research labs. It became a part of NASA in 1958. While its engineers concentrated on aeronautical research and development, its Space Task Group managed the Mercury program and its staff would go on to oversee the Lunar Orbiter and Viking space probes as well as contribute to several other space missions.

LEWIS RESEARCH CENTER was named for George W. Lewis, the long-time director of research for the NACA, and was established in Cleveland, Ohio, in 1941 as the Aircraft Engine Research Laboratory. This center, which became a part of NASA on October 1, 1958, continued to concentrate on flight propulsion research, moving into jet and then rocket technology during the 1940s and 1950s.

MARSHALL SPACE FLIGHT CENTER was established in Huntsville, Alabama, in 1960. Named for former five-star general and Secretary of State George C. Marshall, this organization was transferred from the Army's Redstone Arsenal. Under the leadership of Wernher von Braun, it engaged in rocket development, including the family of Saturn launch vehicles and the launch system for the space shuttle.

STENNIS SPACE CENTER, located near Bay Saint Louis, Mississippi, was established in 1963 by engineers from the Marshall Space Flight Center as a test site for Saturn rockets. Originally called the Mississippi Test Facility, it was renamed for Mississippi Senator John C. Stennis in 1988 and continues to provide a unique test capability for large rockets.

WALLOPS STATION in Wallops Island, Virginia, was established by Langley Research Center in 1946 as a place to carry out sounding rocket research.

NASA ART CREDITS

ACKNOWLEDGMENTS

As in any book, numerous debts are incurred over the course of its production. This is especially true of a breathless introductory survey such as this, which of necessity relies almost exclusively on the primary research of others. I acknowledge the support and encouragement of a large number of people associated with the study of aerospace history and want to thank many individuals who materially contributed to the completion of this project. Of course, I would never have taken this project on except for the encouragement and ideas provided by Bert Ulrich, Curator of the NASA Art Program, and by the fine people directing Stewart, Tabori, and Chang. Thanks also go to Dominick A. Pisano and Patricia Jamison Graboske from the National Air and Space Museum, Smithsonian Institution, for their help and support.

In addition, several individuals read all or part of this manuscript or otherwise offered suggestions which helped me more than they will ever know. My thanks are extended to the staff of the NASA History Division: Stephen J. Garber, who offered valuable advice; Lee D. Saegessor, who helped track down materials and correct inconsistencies; M. Louise Alstork, who provided excellent editorial advice, and Nadine Andreassen, who offered invaluable assistance. In addition to these individuals, I wish to acknowledge the following scholars who aided me in a variety of ways to complete this assignment: Tom D. Crouch, Donald C. Elder, Thomas Fuller, Richard P. Hallion, Francis T. Hoban, Nancy M. House, Sylvia K. Kraemer, John M. Logsdon, John L. Loos, Howard E. McCurdy, Pamela E. Mack, John E. Naugle, Arnauld S. Nicogossian, Robert W. Smith, Rick Sturdevant, Joseph N. Tatarewicz, Shirley Thomas, and Joni Wilson. I also wish to thank the authors of the individual articles for their patience and helpfulness. My thanks also go to Mary Kalamaras for her persistence in editing and to Alexandra Childs for seeing this manuscript through publication.

ROGER D. LAUNIUS

FURTHER READING

Atkinson, Joseph D., Jr., and Shafritz, Jay M. *The Real Stuff: A History of the NASA Astronaut Requirements Program.* New York: Praeger, 1985.

Atwill, William D. *Fire and Power: The American Space Program As Postmodern Narrative.* Athens: University of Georgia Press, 1994.

Benson, Charles D. and Faherty, William Barnaby. *Moonport: A History of Apollo Launch Facilities and Operations.* Washington, DC: NASA Special Publication 4204, 1978.

Bilstein, Roger E. *Flight in America: From the Wrights to the Astronauts.* Baltimore, MD: Johns Hopkins University Press, 1984, paperback reprint 1994.

_____. *Orders of Magnitude: A History of the NACA and NASA, 1915-1990.* Washington, DC: NASA Special Publication 4406, 1989.

_____. *Stages to Saturn: A Technological History of the Apollo/Saturn Launch Vehicles.* Washington, DC: NASA Special Publication 4206, 1980.

Bonnet, Roger M., and Manno, Vittorio. *International Cooperation in Space: The Example of the European Space Agency.* Cambridge, MA: Harvard University Press, 1994.

Braun, Wernher von; Ordway, Frederick I., III; and Dooling, Dave. History of Rocketry and Space Travel. New York: Thomas Y. Crowell Co., 1986 ed.

Brooks, Courtney G.; Grimwood, James M.; and Swenson, Loyd S., Jr. *Chariots for Apollo: A History of Manned Lunar Spacecraft.* Washington: NASA Special Publication 4205, 1979.

Bulkeley, Rip. *The Sputnik Crisis and Early United States Space Policy: A Critique of the Historiography of Space.* Bloomington: Indiana University Press, 1991.

Byrnes, Mark E. *Politics and Space: Image Making by NASA.* New York: Praeger, 1994.

CBS News. 10:56:20 PM EDT, 7/20/69: "The Historic Conquest of the Moon as Reported to the American People." New York: Columbia Broadcasting System, 1970.

Chaiken, Andrew. *A Man on the Moon: The Voyages of the Apollo Astronauts.* New York: Viking, 1994.

Collins, Martin J., and Kraemer, Sylvia K. ed. Space: *Discovery and Exploration. Washington, DC:* Hugh Lauter Levin Associates, Inc., for the Smithsonian Institution, 1993.

_____., and Fries, Sylvia D., ed. *A Spacefaring Nation: Perspectives on American Space History and Policy.* Washington, DC: Smithsonian Institution Press, 1991.

Collins, Michael. *Carrying the Fire: An Astronaut's Journeys.* New York: Farrar, Straus and Giroux, 1974.

_____. *Liftoff: The Story of America's Adventure in Space.* New York: Grove Press, 1988.

Compton, W. David, and Benson, Charles D. *Living and Working in Space: A History of Skylab. Washington, DC:* NASA Special Publication 4208, 1983.

_____. *Where No Man Has Gone Before: A History of Apollo Lunar Exploration Missions.* Washington, DC: NASA Special Publication-4214, 1989.

Cooper, Henry S. F. *Before Lift-off: The Making of a Space Shuttle Crew.* Baltimore, MD: Johns Hopkins University Press, 1987.

Corrigan, Grace. *A Journal for Christa: Christa McAuliffe, Teacher in Space.* Lincoln: University of Nebraska Press, 1993.

Cortright, Edgar M., ed. *Apollo Expeditions to the Moon.* Washington, DC: NASA Special Publication 350, 1975.

DeVorkin, David H. *Science with a Vengeance: How the Military Created the US Space Sciences After World War II.* New York: Springer-Verlag, 1992.

Dick, Steven J. T*he Biological Universe: The Twentieth Century Extraterrestrial Life Debate and the Limits of Science.* New York: Cambridge University Press, 1996.

Divine, Robert A. *The Sputnik Challenge: Eisenhower's Response to the Soviet Satellite.* New York: Oxford University Press, 1993.

Emme, Eugene M., ed. *The History of Rocket Technology: Essays on Research, Development, and Utility.* Detroit, MI: Wayne State University Press, 1964.

Ezell, Edward Clinton, and Ezell, Linda Neuman. *The Partnership: A History of the Apollo-Soyuz Test Project.* Washington, DC: NASA Special Publication 4209, 1978.

Fries, Sylvia D. *NASA Engineers and the Age of Apollo.* Washington, DC: NASA Special Publication 4104, 1992.

Glennan, T. Keith. *The Birth of NASA: The Diary of T. Keith Glennan.* Edited by J. D. Hunley. Washington, DC: NASA Special Publication 4105, 1993.

Green, Constance McL., and Lomask, Milton. *Vanguard: A History.* Washington, DC: NASA Special Publication 4202, 1970; rep. ed. Smithsonian Institution Press, 1971.

Gray, Mike. *Angle of Attack: Harrison Storms and the Race to the Moon.* New York: W. W. Norton and Co., 1992.

Hacker, Barton C., and Grimwood, James M. *On Shoulders of Titans: A History of Project Gemini.* Washington, DC: NASA Special Publication-4203, 1977.

Hallion, Richard P., and Crouch, Tom D., eds. *Apollo: Ten Years Since Tranquility Base.* Washington, DC: Smithsonian Institution Press, 1979.

Jenkins, Dennis R. *Space Shuttle: The History of Developing the National Space Transportation System.* Osceola, WI: Motorbooks, Inc., 1993, 1996.

Kauffman, James L. *Selling Outer Space: Kennedy, the Media, and Funding for Project Apollo, 1961-63.* Tuscaloosa: University of Alabama Press, 1994.

Kay, W. D. *Can Democracies Fly in Space? The Challenge of Revitalizing the U.S. Space Program.* Westport, CT: Praeger, 1995.

Koppes, Clayton R. *JPL and the American Space Program: A History of the Jet Propulsion Laboratory.* New Haven, CT: Yale University Press, 1982.

Kosloski, Lillian D. *U.S. Space Gear: Outfitting the Astronaut.* Washington, DC: Smithsonian Institution Press, 1993.

Krug, Linda T. *Presidential Perspectives on Space Exploration: Guiding Metaphors from Eisenhower to Bush.* New York: Praeger, 1991.

Lambright, W. Henry. *Powering Apollo: James E. Webb of NASA.* Baltimore, MD: Johns Hopkins University Press, 1995.

Launius, Roger D. *Frontiers of Space Exploration.* Westport, CT: Greenwood Press, 1998.

_____. NASA: *A History of the U.S. Civil Space Program.* Melbourne, FL: Krieger Pub. Co., 1994.

_____, and McCurdy, Howard E., Eds. *Spaceflight and the Myth of Presidential Leadership. Urbana:* University of Illinois Press, 1997.

Levine, Alan J. *The Missile and Space Race.* New York: Praeger, 1994.

Levine, Arnold S. *Managing NASA in the Apollo Era.* Washington, DC: NASA Special Publication 4102, 1982.

Lewis, Richard S. *The Last Voyage of Challenger.* New York: Columbia University Press, 1988.

_____. *The Voyages of Apollo: The Exploration of the Moon.* New York: Quadrangle, 1974.

_____. *The Voyages of Columbia: The First True Spaceship.* New York: Columbia University Press, 1984.

Link, Mae Mills. *Space Medicine in Project Mercury.* Washington, DC: NASA Special Publication-4003, 1965.

Logsdon, John M. *The Decision to Go to the Moon: Project Apollo and the National Interest.* Cambridge, MA: The MIT Press, 1970.

_____, General Editor. *Exploring the Unknown: Selected Documents in the History of the U.S. Civil Space Program.* 3 vols. Washington, DC: NASA Special Publication 4407, 1995–97.

Lovell, Jim, and Kluger, Jeffrey. *Lost Moon: The Perilous Voyage of Apollo 13.* Boston: Houghton Mifflin Co., 1994.

McCurdy, Howard E. *Inside NASA: High Technology and Organizational Change in the U.S. Space Program.* Baltimore, MD: Johns Hopkins University Press, 1993.

_____. *Space and the American Imagination.* Washington, DC: Smithsonian Institution Press, 1997.

_____. The *Space Station Decision: Incremental Politics and Technological Choice.* Baltimore, MD: Johns Hopkins University Press, 1990.

McDougall, Walter A. *The Heavens and the Earth: A Political History of the Space Age.* New York: Basic Books, 1985.

Mailer, Norman. *Of a Fire on the Moon. Boston:* Little, Brown, 1970. London, Weidenfeld & Nicolson, 1970. New York: New American Library, 1971.

Murray, Charles A., and Cox, Catherine Bly. *Apollo: The Race to the Moon.* New York: Simon and Schuster, 1989.

Neufeld, Michael J. *The Rocket and the Reich: Peenemünde and the Coming of the Ballistic Missile Era.* New York: Free Press, 1995.

Ordway, Frederick I., III, and Lieberman, Randy, Ed. *Blueprint for Space: From Science Fiction to Science Fact.* Washington, DC: Smithsonian Institution Press, 1992.

_____, and Sharpe, Mitchell R. *The Rocket Team.* New York: Crowell, 1979.

Pitts, John A. *The Human Factor: Biomedicine in the Manned Space Program* to 1980. Washington, DC: NASA Special Publication 4213, 1985.

Reeves, Robert. *The Superpower Space Race: An Explosive Rivalry through the Solar System.* New York: Plenum Press, 1994.

Roman, Peter J. *Eisenhower and the Missile Gap.* Ithaca, NY: Cornell University Press, 1995.

Rosholt, Robert L. *An Administrative History of NASA, 1958–1963.* Washington, DC: NASA Special Publication 4101, 1966.

Sagan, Carl. *Cosmos.* New York: Random House, 1980.

______. *Pale Blue Dot: A Vision of the Human Future in Space.* New York: Random House, 1994.

Seamans, Robert C., Jr. Aiming at Targets: *The Autobiography of Robert C. Seamans, Jr.* Washington, DC: NASA Special Publication 4106, 1996.

Shepard, Alan, and Slayton, Deke. *Moonshot: The Inside Story of America's Race to the Moon.* New York: Turner Publishing, Inc., 1994.

Smith, Robert W. *The Space Telescope: A Study of NASA, Science, Technology, and Politics.* New York: Cambridge University Press, 1989, rev. ed. 1994.

Stuhlinger, Ernst, and Ordway, Frederick I., III. *Wernher von Braun: Crusader for Space.* 2 vols. Malabar, FL: Krieger Pub. Co., 1994.

Swenson, Loyd S., Jr.; Grimwood, James M.; and Alexander, Charles C. *This New Ocean: A History of Project Mercury.* Washington, DC: NASA Special Publication-4201, 1966.

Trento, Joseph J., with reporting and editing by Susan B. Trento. *Prescription for Disaster: From the Glory of Apollo to the Betrayal of the Shuttle.* New York: Crown Publishers, 1987.

Vaughan, Diane. *The Challenger Launch Decision: Risky Technology, Culture, and Deviance at NASA.* Chicago: University of Chicago Press, 1996.

Wilford, John Noble. *Mars Beckons: The Mysteries, the Challenges, the Expectations of Our Next Great Adventure in Space.* New York: Alfred A. Knopf, 1990.

Wilhelms, Don E. *To a Rocky Moon: A Geologist's History of Lunar Exploration.* Tucson: University of Arizona Press, 1993.

Winter, Frank H. *Prelude to the Space Age: The Rocket Societies, 1924-1940.* Washington, DC: Smithsonian Institution Press, 1983.

______. *Rockets into Space.* Cambridge, MA: Harvard University Press, 1990.

Wolfe, Tom. *The Right Stuff.* New York: Farrar, Straus, & Giroux, 1979.

INDEX